HIDDEN ORDER

HIDDEN ORDER

How Adaptation Builds Complexity

John H. Holland

§

HELIX BOOKS

♦♦ *Addison-Wesley Publishing Company*

Reading, Massachusetts · Menlo Park, California · New York
Don Mills, Ontario · Wokingham, England · Amsterdam · Bonn
Sydney · Singapore · Tokyo · Madrid · San Juan
Paris · Seoul · Milan · Mexico City · Taipei

Library of Congress Cataloging-in-Publication Data

Holland, John H. (John Henry), 1929-
    Hidden order : how adaption builds complexity / John H. Holland.
       p.   cm.—(Helix books)
    Includes bibliographical references and index.
    ISBN 0-201-40793-0
    1. Adaptive control systems—Mathematical models.   2. Adaptation
(Biology)—Mathematical models. I. Title. II. Series: Helix books.
TJ217.H64   1995
003.7—dc20   95-20042
                                          CIP

Jacket design by Lynne Reed
Text design by Diane Levy
Set in 10½-point Bembo by Pagesetters, Incorporated
1 2 3 4 5 6 7 8 9 10-DOH-98979695
First printing, July 1995

The Ulam Lecture Series

THIS BOOK IS the first of a continuing series of books based on the Stanislaw M. Ulam Memorial Lectures given at the Santa Fe Institute in Santa Fe, New Mexico. These annual invited lectures, sponsored jointly by the Institute and Addison-Wesley Publishing Company, are dedicated to the memory of Stanislaw Ulam, a great mathematician from the now legendary Polish School of Mathematics. Ulam came to the Institute for Advanced Study in 1935, worked at Harvard, the University of Wisconsin, and much later at the University of Colorado. Most importantly, he joined the Los Alamos National Laboratory in its founding year and was an intellectual force and inspiration there from 1944 until his death in 1984, fostering a perhaps uniquely intense interaction between mathematics and science.

As a mathematician, Stanislaw Ulam held his own with the likes of Kuratowski, Mazur, Banach, von Neumann, and Erdos and his work ranged widely over mathematics. But he was much more, a scientist with a variety of interests who worked with many of the great scientists of the age. Among the topics on which he and his collaborators did founding work were the Monte Carlo method, computer simulations of nonlinear dynamical systems, thermonuclear processes, space propulsion, metrics for biological sequences, cellular automata, and much more. The list of his scientific friends and collaborators

includes many of the greatest minds of the twentieth century. Stan
Ulam's interests in science knew no artificial boundaries; his approach
was truly interdisciplinary. As Françoise Ulam has said, "Stan was a
sort of one-man Santa Fe Institute." He would have loved the
Institute's interdisciplinary, interactive atmosphere and would have
contributed much. It is our loss that he died within a few days of the
official founding of SFI.

—L. M. SIMMONS, JR.
VICE PRESIDENT FOR ACADEMIC AFFAIRS
SANTA FE INSTITUTE

IT IS A GREAT HONOR for me to be here tonight to celebrate with you the tenth anniversary of the creation of the Santa Fe Institute. I want to express all my thanks and appreciation to its founder, my good friend George Cowan, to its leaders, Ed Knapp and Mike Simmons, and to all the other persons who were involved in creating this new series of institute lectures as a memorial to my late husband.

For those of you who did not know Stan Ulam, let me just say a word or two about him.

In a sense, Stan was a sort of one-man SFI because of the interdisciplinary nature of his pursuits. But that was so long ago that the expression had hardly been coined yet.

Were he alive today, he would love SFI's unstructured informality, for he had very little use for all the trappings of bureaucracy and authority. He loved to claim that the only committee he ever served on was the Wine-Tasting Committee of the Junior Fellows at Harvard.

At Los Alamos, he and Carson Mark, the Theoretical Division leader, once confounded the Lab by creating and circulating an official interoffice memo that listed the numbers from one to one hundred in alphabetical order "for quick and easy reference."

When he was promoted to group leader he delighted in the fact that he was a group leader of one, namely himself; for at first he was the only member of his group.

Stan, you see, was a very playful man. And he never considered thinking "work" but rather "play," as in playing with mathematical ideas or inventing mathematical games. He also took great delight in playing with words.

The clever title of tonight's lecture, "Complexity Made Simple," would please him very much, I think, because it is the kind of paradox he liked. So without further ado I will yield the floor to the next speaker, so we can listen to John Holland explain to us in simple terms what complex systems are all about.

—FRANÇOISE ULAM
AT THE INAUGURATION OF THE ULAM LECTURES

Sometimes I feel that a more rational explanation for all that has happened during my lifetime is that I am still only thirteen years old, reading Jules Verne or H. G. Wells, and have fallen asleep.

—STANISLAW ULAM
Adventures of a Mathematician (1976)

The man who had the highest record of accurate guesses in mathematics, the man who could beat engineers at their game, who could size up characters and events in a flash, was a member of an all-but-extinct profession, the profession of prophet.

—GIAN-CARLO ROTA
IN MEMORIAM: STANISLAW ULAM
Notices of the American Mathematical Society (1989)

. . . Its point of view? One
with the twister in vista glide,
 and the cricket in the ditch,
  with river rain, and turbines' trace:
  within the latent
marrow of the egg, the amber
  waves—is where

      its vantage lies.
  Entering the tornado's core,
   entering the cricket waltzed by storm—
to confiscate the shifting give
  and represent the with—
   out which.

    —ALICE FULTON

Contents

Preface

IN THE FALL of 1993, Ed Knapp, president of the Santa Fe Institute, and Jack Repcheck, then editor-in-chief of Helix Books at Addison-Wesley, approached me with a request: Would I inaugurate the Ulam Lectures? The series was to be an annual event, honoring the renowned twentieth-century polymath Stanislaw Ulam. The lectures were to be aimed at a general, science-interested audience, and they were to be expanded into a book so that there would be a permanent record. Although I am quite active in institute affairs, the request came as a complete surprise.

At first I was apprehensive because the time was short—the lectures were to be given sometime in the first half of 1994 and a publishable manuscript was due at the end of that summer. But there were several incentives.

At the top of the list was my long admiration of Stan Ulam's work. When I was a student, there were a few contemporary scientists whose work and abilities I particularly admired: John von Neumann, Ronald Fisher, and Robert Oppenheimer. In pursuing the many facets of von Neumann's work, I repeatedly came across the name Stanislaw Ulam in contexts close to my main interests. So I began to look into his work. That was the beginning of an increasing affinity for Ulam's approach to science, an affinity considerably enhanced when I read his 1976 book,

Adventures of a Mathematician. (There was also a period when I was convinced that Stanislaw Lem, a Polish science fiction writer of Wellsian stature, was Stan Ulam's pen name.) When I was offered a year in Los Alamos as Ulam Scholar, the chance to get to know the place that had supplied the setting for much of his career played its part in my acceptance. It was the only time I ever met him. Later, when Françoise Ulam donated Stan's private library to the Santa Fe Institute, I was delighted to see how many books my own library held in common with his. *Ma bibliothèque, c'est moi.*

Those same thoughts strongly influenced my decision to take on the present commitment. As I began thinking seriously about what would be entailed, I began to see the lectures as an unusual opportunity to make explicit the pattern underlying the intuitions and ideas that had been guiding my research since graduate student days. Writing for a more general audience would force me to strive for bridges and the kind of coherent overview not usually forced on technical work. That was a challenge difficult to ignore.

Then there was my aerie, just completed on the far nothern shore of Lake Michigan and designed for this kind of effort. What a grand way to initiate it! There were other reasons, too, including a nice financial inducement, but they played a lesser role in the decision.

This book centers on an area that has received considerable notice recently: complexity. Stan Ulam made many focusing remarks about complexity, using the word repeatedly and carefully, long before the subject was popular or even named. Many of the themes here are prefigured in Ulam's comments. In writing the book, I have concentrated on that aspect of complexity that centers on adaptation, an area now known as the study of *complex adaptive systems* (*cas*). It is my own bias, as you will see from the book's content, that adaptation gives rise to a kind of complexity that greatly hinders our attempts to solve some of the most important problems currently posed by our world.

I have not tried for a comprehensive review of work relevant to *cas*, nor have I tried to critique other approaches. Instead, I have put all of my effort into producing a single, coherent view of a nascent discipline. The resulting volume is certainly idiosyncratic, though I think many of my colleagues at the Santa Fe Institute would agree with many parts of it. Along with trying to provide an orderly overview, I have also tried to give some feeling for the way a scientist attempts to develop a new discipline. "Doing science," particularly the synthesis of disparate ideas, is not as arcane as it is often made out to be. Discipline and taste play a vital role, but the activity is familiar to anyone who has made some effort to be creative.

The views presented here have been honed through regular interaction with two groups that have played a central role in my scientific development. My longest affiliation is with the BACH group at the University of Michigan (the current members are Arthur Burks, Robert Axelrod, Michael Cohen, John Holland, Carl Simon, and Rick Riolo). We have been meeting regularly for more than two decades, and four of the current members have been active participants from the start. BACH is highly interdisciplinary—five departments are represented—and highly informal, appearing on no roster or organization chart in the university. Almost every idea in this book has been "batted around" before the BACH group at one time or another.

The second group that has played a major role in my outlook is, of course, the Santa Fe Institute (SFI). Though my association with SFI is more recent than my association with BACH, it is no less important to me. The institute encourages deep interdisciplinary science more effectively than any other organization I have encountered. As a graduate student, I thought that the kind of interaction the institute encourages would be the "bread and butter," or at least the "frosting on the cake," of a scientist's activity. Alas, that is rarely the fact. In a university, much time is taken up by advisory and administrative committees, grant seeking and grant administration, negotiation of interdepartmental and intercollegiate cooperation for proposed interdisciplinary activities, and so on. Add in the primary duties of teaching and publication, and there is precious little time for extended interdisciplinary explorations. SFI consistently provides what is hard to come by in a university setting, the opportunity for sustained interdisciplinary research. The institute came into being through the insights and careful organizational work of George Cowan and was soon augmented by an advisory board of scientists who were good at listening as well as presenting. That story has been told elsewhere, by Mitch Waldrop in his 1992 book, *Complexity*, so I will not repeat it here. Suffice it to say that SFI provides a scientific environment that comes very close to the ideal I held as a student.

The event that ultimately led to my association with SFI was an invitation from Doyne Farmer to deliver a talk at one of the annual conferences of the Center of Nonlinear Studies at the Los Alamos National Laboratory. It was that conference that first introduced me to Murray Gell-Mann. He later invited me to join the SFI advisory board, which in turn led to a sustained interaction. That connection provided me with a friend and critic par excellence. In trying to meet Murray's standards for explanation, I have found myself repeatedly refining my ideas about *cas*, attempting to strengthen their foundation and broaden their applicability. It has been an exhilarating exercise that is by no means concluded. Of course, Murray is not the only person at SFI who has influenced my work—the list is quite long and for the most part is chronicled in Waldrop's book—but I think it is fair to say that no other interactions have challenged me to the same degree.

The National Science Foundation has consistently supported my work over several decades, first when I was part of the Logic of Computers Group at the University of Michigan, with Arthur Burks as principal investigator; then, in later years, when Burks and I became co–principal investigators. When I was a young faculty member at Michigan, it was Art Burks who used his prestige to enable me to go down paths that were not part of the traditional university regime. He has been a close friend and mentor for almost forty years.

The MacArthur Foundation recently elected me a MacArthur Fellow. It was Murray Gell-Mann and his wife, Marcia, who informed me of the honor. (And yes, I was in the shower when the call came.) There is really no way to describe the feeling of freedom and elation that accompanies such an award. For good or for ill, the financial security it conveys has encouraged me to take ever-riskier steps in research. Decisions about longer-term projects with uncertain return, such as this book, are much easier.

I would be more than remiss if I failed to mention Françoise Ulam's introduction to the Ulam Lectures. You can read her words at the beginning of this book—but words on paper cannot convey the grace of its delivery. I first met Françoise at the reception preceding the lectures, where we had time for an extended conversation. Her charm and intelligence immediately created a niche of liveliness and warmth in a room full of conversations. It is easy to see why she influenced all aspects of Stan Ulam's research and life, a point he made repeatedly in his autobiography.

I have left my wife, Maurita, for the end of this preface. She has been my constant proxy for the intelligent, science-interested layperson. She has helped in many ways, over and beyond supplying support and encouragement. Early on, it was Maurita who suggested the name "Echo" for the *cas* models described in this book. She has read the chapters that follow many times. Perhaps more willing than the average reader to accept my good intent, in all other respects she has been an effective, unbiased critic. Where this book shows some piece of clarity or untrammeled phrasing, it is likely to be because of her suggestions.

—JOHN HOLLAND
FRIDHEM
GULLIVER, MICHIGAN
APRIL 1995

· 1 ·

Basic Elements

ON AN ORDINARY DAY in New York City, Eleanor Petersson goes to her favorite specialty store to pick up a jar of pickled herring. She fully expects the herring to be there. Indeed, New Yorkers of all kinds consume vast stocks of food of all kinds, with hardly a worry about continued supply. This is not just some New Yorker persuasion; the inhabitants of Paris and Delhi and Shanghai and Tokyo expect the same. It's a sort of magic that everywhere is taken for granted. Yet these cities have no central planning commissions that solve the problems of purchasing and distributing supplies. Nor do they maintain large reserves to buffer fluctuations; their food would last less than a week or two if the daily arrivals were cut off. How do these cities avoid devastating swings between shortage and glut, year after year, decade after decade?

The mystery deepens when we observe the kaleidoscopic nature of large cities. Buyers, sellers, administrations, streets, bridges, and buildings are always changing, so that a city's coherence is somehow imposed on a perpetual flux of people and structures. Like the standing wave in front of a rock in a fast-moving stream, a city is a pattern in time. No single constituent remains in place, but the city persists. To enlarge on the previous question: What enables cities to retain their coherence despite continual disruptions and a lack of central planning?

There are some standard answers to this question, but they really do not resolve the mystery. It is suggestive to say that Adam Smith's "invisible hand," or commerce, or custom, maintains the city's coherence, but we still are left asking How?

Other patterns in time exhibit similar riddles. For instance, if we shift to the microscopic level, we find another community every bit as complicated as New York City. The human immune system is a community made up of large numbers of highly mobile units called *antibodies* that continually repel or destroy an ever-changing cast of invaders called *antigens*. The invaders—primarily biochemicals, bacteria, and viruses—come in endless varieties, as different from one another as snowflakes. Because of this variety, and because new invaders are always appearing, the immune system cannot simply list all possible invaders. It must change or adapt (Latin, "to fit") its antibodies to new invaders as they appear, never settling to a fixed configuration. Despite its protean nature, the immune system maintains an impressive coherence. Indeed, your immune system is coherent enough to provide a satisfactory scientific definition of your *identity*. It is so good at distinguishing you from the rest of the world that it will reject cells from any other human. As a result, a skin graft even from a sibling requires extraordinary measures.

How does the immune system develop its exquisite sense of identity, and what makes that identity vulnerable? How does an immune disease such as AIDS manage to destroy the identity? We can say that the identifications, and the misidentifications, are a product of "adaptation," but the "how" of this adaptive process is far from obvious.

It is more than an academic quest to try to understand the persistence and operation of these two complex communities. Pressing problems, such as prevention of inner-city decay and control of diseases such as AIDS, turn on this understanding. Once we look in this direction, we see that there are other complex systems that pose similar questions, and they too present troubling, long-range problems.

Consider the mammalian central nervous system (CNS). Like the immune system, the CNS consists of a large number of component cells, called neurons, that occur in a wide range of forms. Even a simple CNS consists of hundreds of millions of neurons, of hundreds of types, and each neuron directly contacts hundreds, even thousands, of other neurons to form a complex network. Pulses of energy flash over this network, producing Sherrington's (1951) "enchanted loom." This network is similar to the immune system, with an aggregate emergent identity that *learns* speedily and with great facility. Though the activity of an individual neuron can be complex, it is clear that the behavior of the CNS aggregate identity is much more complex than the sum of these individual activities. The behavior of the central nervous system depends on the *interactions* much more than the actions. The sheer number of interactions—hundreds of millions of neurons, each undergoing thousands of simultaneous interactions in thousandths of a second—takes us well beyond any of our experience with machines. The most sophisticated computer, in comparison, seems little more than an automated abacus. The myriad interactions, modified by learned changes, yield the unique ability of canids, felines, primates, and other mammals to anticipate the consequences of their actions by modeling their worlds.

After more than a century of intensive effort, we still cannot model many basic capabilities of the CNS. We cannot model its ability to parse complex unfamiliar scenes into familiar elements, let alone its ability to construct experience-based internal models. The relation between the distributed, diverse CNS and the phenomenon we call consciousness is largely unknown, a mystery that leaves us with few guidelines for the treatment of mental diseases.

Ecosystems share many of the features and puzzles presented by an immune system or a CNS. They exhibit the same overwhelming diversity. We have yet to assay the range of organisms present in a cubic meter of temperate-zone soil, let alone the incredible arrays of species in a tropical forest. Ecosystems are continually in flux and exhibit a wondrous panoply of interactions such as mutualism, parasitism, biological arms races, and mimicry (more about these later). Matter, energy, and information are shunted around in complex cycles. Once again, the whole is more than the sum of its parts. Even when we have a catalog of the activities of most of the participating species, we are far from understanding the effect of changes in the ecosystem. For example, the stupendous richness of the tropical forest biome contrasts with the poverty of its soil. The forest can only maintain its diversity through a complex set of interactions that recycle sparse nutrients through the system, over and over again.

Until we have an understanding of these complicated, changing interactions, our attempts to balance extraction of ecosystem resources against sustainability will remain at best naive, at worst disastrous. We, as humans, have become so numerous that we perforce extensively modify ecological interactions, with only vague ideas of longer-range effects. Yet our well-being, even our survival, depends on our being able to use these systems without destroying them. Attempts to turn tropical forest into farmland, or to fish the Grand Banks "efficiently," are only symptoms of a problem that year by year becomes more serious.

Many other complex systems show coherence in the face of change. But we can already begin to extract some of the commonalities, and we will later examine additional systems in this light. We can see, for instance, that the coherence and persistence of each system depend on extensive interactions, the aggregation of diverse elements, and adaptation or learning. We have also noted that several perplexing problems of contemporary society—inner-city decay, AIDS, mental disease and deterioration, biological sustainability—are likely to persist until we develop an understanding of the dynamics of these systems. We will see that economies, the Internet, and developing embryos offer similar challenges—trade balances, computer viruses, and birth defects, for example—and we will encounter still others.

Even though these complex systems differ in detail, the question of coherence under change is the central enigma for each. This common factor is so important that at the Santa Fe Institute we collect these systems under a common heading, referring to them as *complex adaptive systems* (*cas*). This is more than terminology. It signals our intuition that general principles rule *cas* behavior, principles that point to ways of solving the attendant problems.

Our quest is to extract these general principles. The quest is new, so this book can only begin to map the territory. And much of that map will consist of terra incognita and legends such as "monsters exist here." Nevertheless, we have come far enough to do more than make casual comparisons. In this first chapter, we can observe some of the prominent landmarks and we can estimate what kinds of apparatus will be needed to come to a broad understanding of complex adaptive systems.

Objectives

The purpose of this book is to explore the ways in which our intuitions about *cas* can be transformed into a deeper understanding. Theory is crucial. Serendipity may occasionally yield insight, but is unlikely to be a frequent visitor. Without theory, we make endless forays into uncharted badlands. With theory, we can separate fundamental characteristics from fascinating idiosyncrasies and incidental features. Theory supplies landmarks and guideposts, and we begin to know what to observe and where to act.

One specific piece of understanding that theory could supply is a more principled way of locating "lever points" in *cas*. Many *cas* have the property that a small input can produce major predictable, directed changes—an amplifier effect. A familiar example is a vaccine. An infection into our bloodstream of a small amount of an incapacitated antigen, say the measles virus, can stimulate the immune system to produce enough antibodies to make us completely immune to the disease. The vaccine "levers" the immune system into learning about the disease, saving the costly, uncomfortable procedure of learning about the disease "on line." We know of other lever points in other *cas*, but to date we have no comprehensive method of searching them out. Theory is our best hope of finding such a method.

The task of formulating theory for *cas* is more than usually difficult because the behavior of a whole *cas* is more than a simple sum of the behaviors of its parts; *cas* abound in nonlinearities (more about this shortly). Nonlinearities mean that our most useful tools for generalizing observations into theory—trend analysis, determination of equilibria, sample means, and so on—are badly blunted. The best way to compensate for this loss is to make cross-disciplinary comparisons of *cas*, in hopes of extracting common characteristics. With patience and insight we can shape those characteristics into building blocks for a general theory. Cross-comparisons provide another advantage: characteristics that are subtle and hard to extract from one system can be salient and easy to examine in another. This chapter is about seven characteristics that cross-disciplinary comparisons suggest are central to a broad understanding of *cas*. Subsequent chapters will weave these characteristics into the elements of a theory.

Agents, Meta-Agents, and Adaptation

Before going on to a description of the characteristics themselves, I should say something more about the general setting. *Cas* are, without exception, made up of large numbers of active elements that, from the examples we've seen, are diverse in both form and capability (see Figure 1.1). Think of the great array of firms in New York City or the exquisitely tuned antibodies in the immune system. To refer to active elements

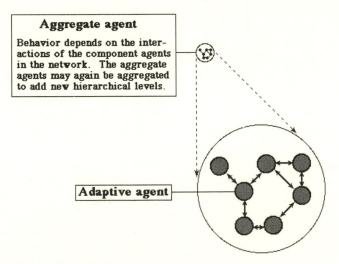

Figure 1.1　A Complex Adaptive System.

without invoking specific contexts, I have borrowed the term *agent* from economics. The term is descriptive but avoids preconceptions.

If we are to understand the interactions of large numbers of agents, we must first be able to describe the capabilities of individual agents. It is useful to think of an agent's behavior as determined by a collection of rules. Stimulus-response rules are typical and simple: IF stimulus *s* occurs THEN give response *r*. IF the market goes down THEN sell. IF the car has a flat tire THEN get the jack. And so on. To define the set of stimulus-response rules possible for a given agent, we must describe the stimuli that agent can receive and the responses it can give (see Figure 1.2).

IF **STIMULUS** **THEN** **RESPONSE**

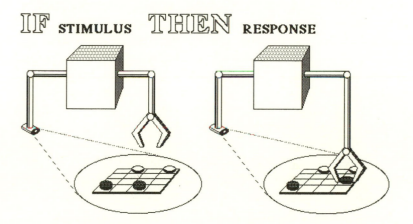

PERFORMANCE (A SUCCESSION OF S-R EVENTS)

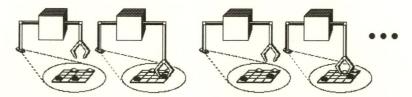

Figure 1.2   A Rule-Based Agent.

Though stimulus-response rules are limited in scope, there are simple ways of expanding that scope. Indeed, with minor changes, the scope can be enlarged sufficiently that clusters of rules can generate any behavior that can be computationally described. In the definition of these rules, our intent is *not* to claim that we can locate the rules explicitly in the real agents. Rules are simply a convenient way to describe agent strategies. In the next chapter I will say more about this rule-based approach to agent behavior; for now, let us treat it as a descriptive device.

A major part of the modeling effort for any *cas*, then, goes into selecting and representing stimuli and responses, because the behaviors and strategies of the component agents are determined thereby. For agents in the central nervous system (neurons), the stimuli could be pulses arriving at each neuron's surface, and the responses could be the outgoing pulses. For agents in the immune system (antibodies), the stimuli could be molecular configurations on the surface of the invading antigens, and the responses could be differing adhesions to the antigen surface. For agents in an economy (firms), the stimuli could be raw materials and money, and the responses could be goods produced. We could make similar selections for other *cas*. The "could" in each case is relevant because other selections are possible. Different selections emphasize different aspects of the *cas*, yielding different models. This is not so much a matter of correct or incorrect (though models can be poorly conceived) as it is a matter of what questions are being investigated.

Once we specify the range of possible stimuli and the set of allowed responses for a given agent, we have determined the kinds of rules that agent can have. Then, by looking at these rules acting in sequence, we arrive at the behaviors open to the agent. It is at this point that learning or adaptation enters. In setting up a list of basic elements, we might think it natural to put "adaptation" at the head of the list, because adaptation is the sine qua non of *cas*. But adaptation is such a broad topic that it encompasses almost everything else in this book. The present chapter centers on the more specific characteristics of *cas,* so I will only say a few words about adaptation here and provide a more careful discussion in the next chapter.

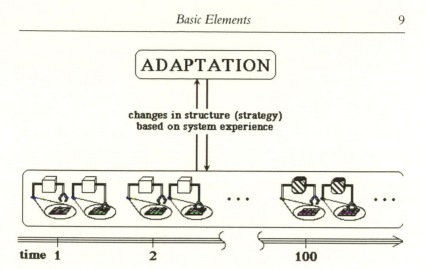

SYSTEM	MODIFICATION TIME
central nervous system	seconds to hours
immune system	hours to days
business firm	months to years
species	days to centuries
ecosystem	years to millennia

Figure 1.3   Adaptation and Learning.

Adaptation, in biological usage, is the process whereby an organism fits itself to its environment. Roughly, experience guides changes in the organism's structure so that as time passes the organism makes better use of its environment for its own ends (see Figure 1.3). Here we expand the term's range to include learning and related processes. With this extension, adaptation applies to all *cas* agents, despite the different timescales of different *cas* processes. And indeed, the timescales do vary. Adaptive changes in individual neurons in the nervous system take place over an interval that ranges from seconds to hours; adaptive changes in the immune system require hours to days; adaptive changes in a business firm take months to years; adaptive changes in an ecosystem take years to millennia or more. Still, the mechanisms involved in all these cases have much in common, once time is factored out. There is a general framework that supports this extended use of the term (see Holland, 1992), but we do not need that level of detail just now. Parts of that framework will be introduced as needed throughout the book.

Overall, then, we will view *cas* as systems composed of interacting agents described in terms of rules. These agents adapt by changing their rules as experience accumulates. In *cas*, a major part of the environment of any given adaptive agent consists of other adaptive agents, so that a portion of any agent's efforts at adaptation is spent adapting to other adaptive agents. This one feature is a major source of the complex temporal patterns that *cas* generate. To understand *cas* we must understand these ever-changing patterns. The rest of this book is devoted to developing such an understanding by filling in this rough sketch, adding detail, content, and relevant pieces of theory. Now to our seven basics.

Seven Basics

The seven basics consist of four properties and three mechanisms that are common to all *cas*. They are not the only basics that could be selected from a list of common characteristics; the selection process is, in part, a matter of taste. Still, all the other candidates of which I am aware can be "derived" from appropriate combinations of these seven.

In presenting the basics, I have ordered them in a way that emphasizes their interrelations rather than grouping them into properties and mechanisms.

AGGREGATION (*PROPERTY*)

Aggregation enters into the study of *cas* in two senses. The first refers to a standard way of simplifying complex systems. We aggregate similar things into categories—trees, cars, banks—and then treat them as equivalent. Humans analyze familiar scenes in this way with the greatest of ease. Not too surprisingly, the categories we choose are reusable; we can almost always decompose novel scenes into familiar categories. And we can generate scenes we have never seen by recombining the categories—much as the griffin, chimera, and harpy of a medieval bestiary are formed by recombining familiar animal parts in new ways.

Aggregation in this sense is one of our chief techniques for constructing models. We decide which details are irrelevant for the questions of interest and proceed to ignore them. This has the effect of collecting into a category things that differ only in the abandoned details; the category becomes a building block for the model. Modeling, it should be clear, is an art form. It depends on the experience and taste of the modeler. In this it is much like cartooning, especially political cartooning. The modeler (cartoonist) must decide which features to make salient (exaggerate), and which features to eliminate (avoid), in order to answer the questions (make the political point).

The second sense of aggregation is closely related to the first, but it is more a matter of *what cas do*, rather than *how* we model them. It concerns the emergence of complex large-scale behaviors from the aggregate interactions of less complex agents. An ant nest serves as a familiar example. The individual ant has a highly stereotyped behavior, and it almost always dies when circumstances do not fit the stereotype. On the other hand, the ant aggregate—the ant nest—is highly adaptive, surviving over long periods in the face of a wide range of hazards. It is much like an intelligent organism constructed of relatively unintelligent parts. Douglas Hofstadter's wonderful chapter "Ant Fugue" in his 1979 book makes this point better than anything else I have read. In it the ant nest provides a comprehensible version of more spectacular emergent phenomena, such as the intelligence of large numbers of interconnected neurons, or the identity provided by a diverse array of antibodies, or the spectacular coordination of an organism made of myriad cell types, or even the coherence and persistence of a large city.

Aggregates so formed can in turn act as agents at a higher level— *meta-agents*. The interactions of these meta-agents are often best described in terms of their aggregate (first sense) properties (see Figure 1.4). Thus we speak of the gross domestic product of an economy, or the identity provided by the immune system, or the behavior of a nervous system. Meta-agents can, of course, aggregate (second sense) in turn to yield meta-meta-agents. When this process is repeated several times, we get the hierarchical organization so typical of *cas*.

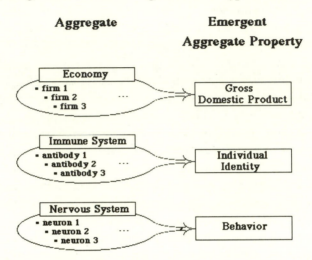

Figure 1.4   Aggregation and Aggregate Properties.

Aggregation in the second sense is indeed a basic characteristic of all *cas*, and the emergent phenomena that result are the most enigmatic aspect of *cas*. The study of *cas* turns on our ability to discern the mechanisms that enable simple agents to form highly adaptive aggregates. What kind of "boundaries" demarcate these adaptive aggregates? How are the agent interactions within these boundaries directed and coordinated? How do the contained interactions generate behaviors that transcend the behaviors of the component agents? We must be able to answer such questions if we are to resolve the mysteries, and the difficulties, that attend *cas*.

TAGGING (*MECHANISM*)

There is a mechanism that consistently facilitates the formation of aggregates—a mechanism that in this book will go by the name *tagging*.

The most familiar example is a banner or flag that is used to rally members of an army or people of similar political persuasion. A more operational version of a tag, in these days of Internet, is the header on a message that knits together members of a bulletin board or conference group. Still more operational are the "active sites" that enable antibodies to attach themselves to antigens. The sophistication of this particular version of tagging is well described in Edelman's (1988) discussion of cell adhesion molecules. We can continue with the visual patterns and pheromones that facilitate selective mating in animals, and the trademarks, logos, and icons that facilitate commercial interaction (see Figure 1.5). It soon becomes apparent that tagging is a pervasive mechanism for aggregation and boundary formation in *cas*.

When we closely examine different instances of tagging, we see there is a common feature: *cas* use tags to manipulate symmetries. Because symmetries are common, we often use them in perceiving or modeling our day-to-day world, sometimes quite unconsciously. They enable us to ignore certain details, while directing our attention to others. Weyl (1952) gives a rich exposition of this point. The classic example of a full-blown symmetry is a perfect sphere, say the white cue ball in billiards. A cue ball exhibits complete rotational symmetry, so that rotation in any direction produces no noticeable change. If we put a stripe around the cue ball's "equator," turning it into one of the other billiard balls, we break the symmetry, allowing us to distinguish the previously indistinguishable. For example, if we start the striped ball spinning, we can easily observe whether or not the ball's axis of rotation defines the equator marked out by the stripe. Most rotations produce noticeable changes, except for those around the axis that defines the cue ball's equator. That is, some symmetries are broken and others remain. In general, tags enable us to observe and act on properties previously hidden by symmetries.

To carry the example a bit further, consider a *set* of cue balls in rapid motion on a billiard table, say after a strong "break." We cannot distinguish the individual cue balls unless we keep a careful record of their trajectories. But again, we can break the symmetry via a tag. If we

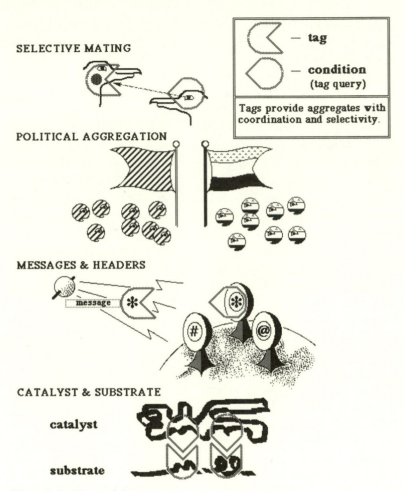

Figure 1.5   Tags and Aggregates.

put a striped cue ball in with the other cue balls, we can easily track it despite its motion.

Tags are a pervasive feature of *cas* because they facilitate selective interaction. They allow agents to select among agents or objects that would otherwise be indistinguishable. Well-established tag-based interactions provide a sound basis for filtering, specialization, and cooperation. This, in turn, leads to the emergence of meta-agents and organizations that persist even though their components are continually changing. Ultimately, tags are the mechanism behind hierarchical organization—the agent / meta-agent / meta-meta-agent / . . . organization so common in *cas*. We'll see many examples of the origin and intervention of tags as we go along.

NONLINEARITY (*PROPERTY*)

It is little known outside the world of mathematics that most of our mathematical tools, from simple arithmetic through differential calculus to algebraic topology, rely on the assumption of linearity. Roughly, linearity means that we can get a value for the whole by adding up the values of its parts. More carefully, a *function* is *linear* if the value of the function, for any set of values assigned to its arguments, is simply a weighted sum of those values. The function $3x + 5y + z$, for example, is linear.

We say some numerical property of a system is linear, relative to numerical values assigned to its parts, if the property is a linear function of those values. Consider, for example, the fuel consumption c of a plane as a function of its velocity v and its altitude x. Given suitable units for fuel consumption, altitude, and velocity, we might be able to establish that

$$c = (0.5)v + (-0.1)v.$$

Fuel consumption then would be linear in terms of velocity and altitude.

Polls, or project trends, or industrial statistics, all of which employ summation, are only useful if they describe linear properties of the underlying systems. It is so much easier to use mathematics when systems have linear properties that we often expend considerable effort to justify an assumption of linearity. Whole branches of mathematics are devoted to finding linear functions that are reasonable approximations when linearity cannot be directly established. Unfortunately, none of this works well for *cas*. To attempt to study *cas* with these techniques is much like trying to play chess by collecting statistics on the way pieces move in the game.

Let me illustrate the difficulty by starting with one of the simplest *non*linear interactions, that between a predator population and its prey. The model we look at, despite its simple assumptions, does a satisfactory job of describing real data, such as the centuries-long record of lynx-hare interactions derived from the Hudson Bay Company's yearly records of pelt acquisitions. In putting this model together, we sketch a typical procedure for building mathematical models. When we have finished, we'll have a clear example of the complications caused by nonlinearities.

We begin with the commonsense observation that increases in either the predator population or the prey population increase the likelihood of encounters between predator and prey. Symbolically, if U represents the number of predators in a given area, say a square mile, and V represents the number of prey in the same area, then the number of interactions per unit time, say a day, is given by cUV, where c is a constant that reflects the efficiency of the predator (for example, the average rate at which it searches the territory). If $c = 0.5$, $U = 2$, and $V = 10$, then there will be

$$cUV = 0.5(2)(10) = 10 \text{ encounters}$$

per day per square mile. If the number of predators increases by 2, so that $U = 4$, and the number of prey increases by 10, so that $V = 20$, then the number of encounters will be quadrupled to

$$cUV = 0.5(4)(20) = 40 \text{ encounters}$$

per day per square mile. This expression involves a nonlinearity, one of the simplest, because it entails the *product* of two distinct variables instead of their *sum*. That is, the overall predator-prey interaction cannot be obtained merely by adding predator activity to prey activity.

Our next step is to take explicit account of the fact that the populations change over time. Notationally, we let $U(t)$ stand for the population of predators at time t; similarly $V(t)$ stands for the population of prey at time *t*. We augment the predator-prey interaction by allowing for births and for deaths from causes other than predation. Taking the simplest approach, we set a common birthrate *b* for all predators, so that the number of predator births at time *t* is $bU(t)$. Deaths can be handled similarly by using a common death rate *d*, so that the number of predator deaths at time *t* is $dU(t)$.

If we ignore predator-prey interactions for a moment, we arrive at a simple model of the way the population of predators changes over time. The size of the population after one unit of time has elapsed is the population at time *t*, minus the deaths, plus the births, or

$$U(t + 1) = U(t) - dU(t) + bU(t).$$

This equation, with allowances for aging, is the foundation for population projections and such mundane things as life insurance premiums. We use exactly the same argument to get a similar equation for the prey,

$$V(t + 1) = V(t) - d'V(t) + b'V(t),$$

where *b'* and *d'* are the respective birth and death rates for the prey (again without interactions).

To reintroduce the effect of predator-prey interactions, we incorporate the intuitive idea that the predator enhances its well-being each time it catches prey. Ultimately this process exerts a positive effect on the predator's production of offspring. To capture this idea mathematically, introduce another constant *r* that represents the efficiency of transforming captured prey (food) into offspring. More interactions mean more births, so using the interaction rate $cU(t)V(t)$, we get

$$r[cU(t)V(t)]$$

as the enhancement in births because of predator-prey interaction. The population change for predators then becomes

$$U(t + 1) = U(t) - dU(t) + bU(t) + r[cU(t)V(t)].$$

For prey, capture by a predator increases the number of deaths. Using r' to indicate the vulnerability to capture and death during interactions, we obtain a population equation for the prey,

$$V(t + 1) = V(t) - d'V(t) - r'[cU(t)V(t)] + b'V(t).$$

This pair of equations for $U(t + 1)$ and $V(t + 1)$ is a version of the famous Lotka-Volterra model (see Lotka, 1956). Standard ways of simplifying and solving the Lotka-Volterra equations show that, under most conditions, the predator population will go through a series of oscillations between feast and famine, as will the prey population. This prediction is borne out by the Hudson Bay Company's records. In the long run, extensions of such models should help us understand why predator-prey interactions exhibit strong oscillations, whereas the interactions that form a city are typically more stable. For now we are only interested in the effect of nonlinearities on such modeling efforts.

Let's return to the interactive part of the model. The $cU(t)V(t)$ formulation is actually a starting point for many other models, including interactions between atoms or molecules or even billiard balls. To study the effect of nonlinear interactions in the simplest possible setting, we shift back to the billiard balls (see Figure 1.6).

Let's restrict the model to just three "species" of billiard balls: white balls with a red stripe, white balls with an orange stripe, and solid-blue balls. Assume that there are several of each on the table and that they are in random motion—a kind of "big bang" or, better, "big break." Also assume, somewhat fancifully, that the "stripes" sometimes stick to the "solids" when they collide, as if they had dots of Velcro on their surfaces. The earlier formula cUV can now be used to model the rate at which the "stripe/solid compounds" form.

To see this, let's begin with the red-stripe/blue-solid combination. Our U gives the proportion of red-stripes on the table, V gives the proportion of blue-solids, and the constant c now gives a reaction rate that depends on the stickiness of the red-stripe/blue-solid combination. Using $Z(t)$ to represent the proportion of red-stripes stuck to blue-solids at time $t,$ we get a simpler version of the Lotka-Volterra equation,

The simplest models of interactions use *random collisions*

(e.g. atomic, chemical, and predator-prey models)

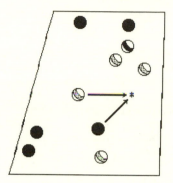

Total number of balls: 10

Proportion of 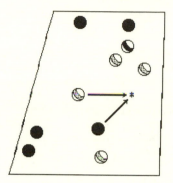: 4/10 = 0.4

Proportion of ●: 5/10 = 0.5

Some collisions produce a compound (a product).

The proportion of collisions resulting in a compound is set by a *reaction rate* using the (nonlinear) equation:

[propor.] x [propor.●] x *rate* = [propor.]
[0.4] x [0.5] x 0.5 = [0.1]

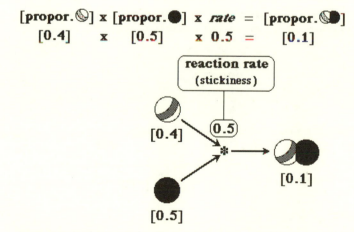

Figure 1.6 A Billiard Ball Model of Interaction.

$$Z(t + 1) = Z(t) + cU(t)V(t).$$

For example, if $Z(t) = 0$, $U(t) = 0.4$, $V(t) = 0.5$, and $c = 0.5$, then after one unit of time the proportion of the red-stripe/blue-solid compound is

$$Z(t + 1) = 0 + 0.5(0.4)(0.5) = 0.1.$$

Different kinds of balls may have different reaction rates:

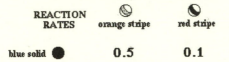

REACTION RATES	orange stripe	red stripe
blue solid	0.5	0.1

Suppose we want to know the proportion of collisions resulting in a stripe-solid compound {◑ and ◐}.

Can we build a simple model by assigning an *aggregate* (average) *reaction rate* to the overall process?

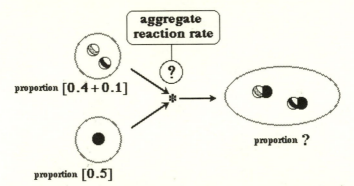

This reaction aggregates the stripes so it uses only the <u>total</u> proportion of stripes {proportion of ◉ + proportion of ◐}.

FOR THE MODEL TO WORK: Two mixes of stripes with the same total must produce the same result.

Figure 1.7 Aggregate Reactions.

We can proceed similarly with the orange-stripes, allowing for the fact that stickiness of the orange-stripes may be different from that of the red-stripes (see Figure 1.7). Let $W(t)$ be the proportion of orange-stripes at time t, let $Y(t)$ be the proportion of the orange-stripe/blue-solid compound at time t, and let c' be the reaction rate determined by the stickiness of orange-stripes. Then, as for the red-stripes, the formula

$$Y(t + 1) = Y(t) + c' W(t) V(t)$$

gives the outcome of the interaction of orange-stripes with blue-solids. If $Y(t) = 0$, $W(t) = 0.1$, and $c' = 0.1$, then

$$Y(t + 1) = 0 + 0.1(0.1)(0.5) = 0.005.$$

We can get the total stripe-solid compound (red-stripe/blue-solid *plus* orange-stripe/blue-solid), $X(t)$, by adding the results of the separate reactions,

$$\begin{aligned} X(t + 1) &= Y(t + 1) + Z(t + 1) \\ &= Y(t) + Z(t) + cU(t) V(t) + c' W(t) V(t). \end{aligned}$$

Using the numerical values given earlier, we obtain

$$X(t + 1) = 0.1 + 0.005 = 0.105.$$

This part of the model is indeed linear—the whole does equal the sum of the parts!

Now suppose we want to simplify the model by aggregating the stripes into a single category. The idea is to calculate the total stripe-solid compound using only the total proportion of stripes on the table. Even when there are only two species of stripes, as in the present case, this aggregation cuts the complication (the number of equations) in half. When there are large numbers of species (as with an ecosystem or a city), aggregation makes the difference between feasibility and infeasibility when it comes to analysis. The simplification occurs because the aggregate equation uses a single variable $S(t)$ for the *total* population of stripes, along with a single reaction coefficient c'', giving the single equation

$$X(t + 1) = X(t) + c''S(t)V(t).$$

There is a problem about the validity of this equation, however. For it to be useful, we must find a coefficient c'' that works for all mixes of stripes.

Under a standard linear approach, we would obtain c'' by averaging the coefficients of the individual stripe-solid interactions. However, this is the point at which the nonlinearities interfere. Consider two different mixes of stripes. In mix 1, the proportion of red-stripes is $U = 0.4$ and the proportion of orange-stripes is $W = 0.1$; in mix 2, the proportion is reversed, so that $U = 0.1$ and $W = 0.4$. In both cases the total number of stripes is $S = U + W = 0.5$. It follows that in both cases the equations for X must give the same answers for the proportion of stripe-solid compound, since all the numbers on the right side are the same. But what actually happens? Do the interactions of the two different mixes really yield the same total proportion of stripe-solid compound?

To check, let's carry out the detailed computation for the two mixes. For mix 1, we have already calculated the result when $X(t) = 0$,

$$X(t + 1) = Y(t + 1) + Z(t + 1) = 0.105.$$

For mix 2, we have

$$\begin{aligned}
X(t + 1) = Y(t + 1) &= Z(t + 1) \\
&= Y(t) + Z(t) + cU(t)V(t) + c'W(t)V(t) \\
&= 0 + 0 + 0.5(0.1)(0.5) + 0.1(0.4)(0.5) \\
&= 0.025 + 0.020 = 0.045.
\end{aligned}$$

And there's the rub. The two mixes produce different compound totals, 0.105 versus 0.045, even though the total number of stripes is the same. No summing or averaging of the reaction coefficients of the aggregate's parts will work, because there is *no* coefficient that will work for both mixes. The nonlinear interactions prevent us from assigning an aggregate reaction rate to the aggregate reaction.

We are now at the end of this particular tale. We've seen that even in the simplest situations nonlinearities can interfere with a linear approach to aggregates. That point holds in general: nonlinear interactions almost always make the behavior of the aggregate more complicated than would be predicted by summing or averaging.

FLOWS (*PROPERTY*)

The idea of flows extends well beyond the movement of fluids. In everyday usage, we speak of the flow of goods into a city or the flow of capital between countries. In more sophisticated contexts, we think of flows over a network of nodes and connectors. The nodes may be factories, and the connectors transport routes for the flow of goods between the factories. Similar {node, connector, resource} triads exist for other *cas*: {nerve cells, nerve cell interconnections, pulses} for the central nervous system; {species, foodweb interactions, biochemicals} for ecosystems; {computer stations, cables, messages} for the electronic Internet; and so on (see Figure 1.8). In general terms, the nodes are processors—*agents*—and the connectors designate the possible interactions. In *cas* the flows through these networks vary over time; moreover, nodes and connections can appear and disappear as the agents adapt or fail to adapt. Thus neither the flows nor the networks are fixed in time. They are patterns that reflect changing adaptations as time elapses and experience accumulates.

Tags almost always define the network by delimiting the critical interactions, the major connections. Tags acquire this role because the adaptive processes that modify *cas* select *for* tags that mediate useful interactions and *against* tags that cause malfunctions. That is, agents with useful tags spread, while agents with malfunctioning tags cease to exist. Later on we will look at this process in some detail.

There are two properties of flows, well known from economics, that are important to all *cas*. The first of these is the *multiplier effect* (see, for

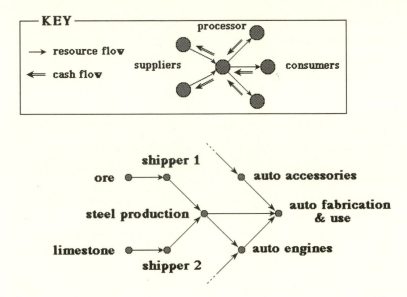

Figure 1.8   Flows.

example, Samuelson, 1948), which occurs if one injects additional resource at some node. Typically this resource is passed from node to node, possibly being transformed along the way, and produces a chain of changes (see Figure 1.9).

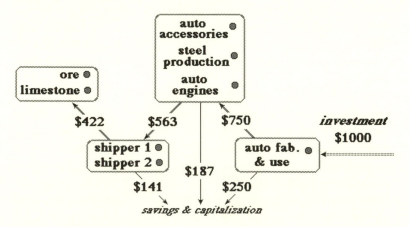

Figure 1.9   Multiplier Effect.

The simplest examples come from economics. When you contract to build a house, you pay the contractor, who pays the tradesmen, who in turn buy food and other commodities, and so on, stage by stage through the economic network. In order to make a simple computation, let's assume that at each stage one-fifth of the new income is saved, and the other four-fifths is paid to the next stage. Then for each dollar you pay, 80 cents will be passed on by the contractor to the tradesmen, who in turn pass on 64 cents, and so on. More generally, a fraction r is passed on at each stage. So at stage 2, a fraction r of the original amount is available. At stage 3, a fraction r of that r is available, so r^2 is available at stage 3. And so it goes for each successive stage. We can calculate the total effect by using the fact that $1 + r + r^2 + r^3 + \ldots = 1/(1 - r)$. In this example, $r = 0.8$, so the total effect is approximately $1/(1 - 0.8) = 5$. That is, the initial effect, your contract, is multiplied by five when we trace its total effect as it passes through the network.

This multiplier effect is a major feature of networks and flows. It arises regardless of the particular nature of the resource, be it goods, money, or messages. It is relevant whenever we want to estimate the effect of some new resource, or the effect of a diversion of some resource over a new path. It is particularly evident when evolutionary changes occur, and it typically jeopardizes long-range predictions based on simple trends.

The second property is the *recycling effect*—the effect of cycles in the networks (see Figure 1.10). This too is most easily explained using an example from economics. Consider a network involving three nodes, say an ore supplier, a steel producer, and a node that stands for auto fabrication and use. For simplicity, we adjust the resource measures (weights) so that one unit of ore produces one unit of steel which in turn produces one unit of automobile. Further, we'll have the steel producer send exactly half its output to the auto fabrication/use node. That is, if the ore supplier ships 1000 units of ore, that will translate through the network to become $0.5(1000) = 500$ units of auto. If we assume that the autos produced are used until they turn into unrecoverable rust, then the return for each 1000 units of ore mined is 500 units of automobile. How do things change if we manage to recycle three-quarters of the steel in autos? Some of the material now goes through a cycle from the fabrication/use node to the shipper through the steel producer and back to the fabrication/use node. Under this arrangement, with the same 1000 units of ore from the miner, steel production settles down at 1600 units output, which in turn yields 800 units of auto at the fabrication/use node. Recycling, with the same raw input, produces more resource at each node.

That recycling can increase output is not particularly surprising, but the overall effect in a network with many cycles can be striking. A tropical rain forest illustrates the point. The soil there is extremely poor because tropical downpours have a leaching effect that quickly moves resources from the soil into the river system. For this reason ordinary agriculture, which does not recycle resources, fares poorly when the tropical forest is cleared. Yet the forest itself is rich in both species and

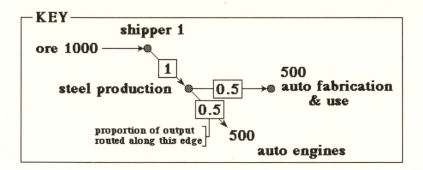

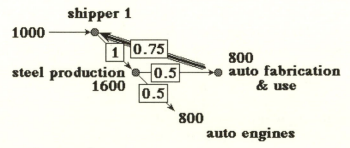

Figure 1.10   Recycling.

numbers of individuals. This state of affairs depends almost entirely on the forest's ability to capture and recycle critical resources. For the forest departs from the simple version of a food web, in which resources are only passed upward to the top predator. Instead, cycle after cycle traps the resources so that they are used and reused before they finally make their way into the river system. The resulting system is so rich that a *single* rain forest tree may harbor over 10,000 distinct species (!) of insects.

DIVERSITY (*PROPERTY*)

In that same tropical rain forest, in addition to the diversity of insects, it is possible to walk half a kilometer without twice encountering the same species of tree. But the rain forest is not an isolated example. The mammalian brain consists of a panoply of neuron morphologies organized into an elaborate hierarchy of nuclei and regions; New York City consists of thousands of distinct kinds of wholesalers and retailers; and so it goes for each *cas* in turn.

This diversity is neither accidental nor random. The persistence of any individual agent, whether organism, neuron, or firm, depends on the context provided by the other agents. Roughly, each kind of agent fills a niche that is defined by the interactions centering on that agent. If we remove one kind of agent from the system, creating a "hole," the system typically responds with a cascade of adaptations resulting in a new agent that "fills the hole." The new agent typically occupies the same niche as the deleted agent and provides most of the missing interactions. This process is akin to the phenomenon called *convergence* in biology. The ichthyosaur of the ancient Triassic seas filled much the same niche as the porpoise in modern seas. Though the ichthyosaur is no kin to the porpoise, it is surprisingly similar in form and habit. It even preyed on cephalopods (squid and octopuses). And here you have another convergence. The eye of a squid exhibits all the parts and complexity of a mammalian eye, yet the two are derived from entirely different tissues. The two eyes fill the same niche in different physiologies, a niche determined by the interactions eyes must provide.

Convergence of a kind also occurs when an established species enters virgin territory. The islands of Hawaii, newly arisen a few million years ago, constituted virgin territory for a pregnant fruit fly (genus *Drosophila*) that drifted or was blown there from elsewhere. Over 600 indigenous species of fruit fly have arisen from that founder. Still more remarkable, these new species fill all sorts of niches that are occupied by very different fly species elsewhere in the world. The ecosystem interactions are largely re-created, although the agents are quite different.

Diversity also arises when the spread of an agent opens a new niche—opportunities for new interactions—that can be exploited by modifications of other agents. Mimicry, a pervasive biological phenomenon, is a good example. In North America the most familiar example of mimicry involves the monarch butterfly (see Figure 1.11). The monarch is marked by a striking orange and black pattern, but it flies quite openly in the fields, unlike most butterflies that flit quickly from cover to cover to avoid predators. The monarch can move so freely because its caterpillar accumulates a bitter alkaloid from the milkweed plant; birds quickly learn that the monarch butterfly induces vomiting. There is a second butterfly, the viceroy, that has a wing pattern almost identical to that of the monarch but it lacks the monarch's bitterness. It mimics the monarch, and thereby gains an important freedom. How can blind chromosomes generate a complicated pattern that mimics the pattern of an entirely different species? It's an important question that we'll look into later, when we have a better foundation. For now we simply note the new niche, and the diversity, provided by the presence of the monarch.

Mimicry exists at every turn in the rain forest. Insects mimic twigs,

Monarch   **Viceroy**

Figure 1.11   Mimicry.

snakes, and even bird splat. Orchids mimic a wide range of pollinators so well that, as in the case of the bee orchid, they induce copulatory movements as a means of covering the insect with pollen. The orchid family itself consists of close to 20,000 species, exhibiting an extraordinary variety of shapes and mechanisms (including pollen-throwing and clasping devices). Each new species opens still newer possibilities for interaction and specialization, with still further increases in diversity.

The diversity of *cas* is a dynamic pattern, often persistent and coherent like the standing wave we alluded to earlier. If you disturb the wave, say with a stick or paddle, the wave quickly repairs itself once the disturbance is removed. Similarly in *cas*, a pattern of interactions disturbed by the extinction of component agents often reasserts itself, though the new agents may differ in detail from the old. There is, however, a crucial difference between the standing wave pattern and *cas* patterns: *cas* patterns evolve. The diversity observed in *cas* is the product of progressive adaptations. Each new adaptation opens the possibility for further interactions and new niches.

What mechanisms enable *cas* to generate and maintain temporal patterns with such diverse components? Answers to this question are pivotal to any deep understanding of *cas*. To have a comprehensive theory, we must answer this question in way that applies to all *cas*. A principle from paleontology applies *mutatis mutandis* here: to understand species, understand their phylogeny.

We can make some progress in comprehending the origins of diversity if we revisit flows in the light of this paleontological principle. Note first that the patterns of interaction familiar from ecology—symbiosis, parasitism, mimicry, biological arms races (see Figure 1.12; Dawkins, 1976, is worth reading on this subject), and so on—are all well described in terms of agent-directed flows of resources. Because these interactions have counterparts in other *cas*, we can extend this observation to them as well. From the earlier discussion of recycling, we know that agents that participate in cyclic flows cause the system to retain resources. The resources so retained can be further exploited—they offer new niches to be exploited by new kinds of agents. Parts of a *cas* that exploit these possibilities, particularly parts that further enhance recycling, will thrive. Parts that fail to do so will lose their resources to those that do. This is natural selection writ large. It is a process that leads to increasing diversity through increasing recycling.

We can further enlarge this view if we add some thoughts about

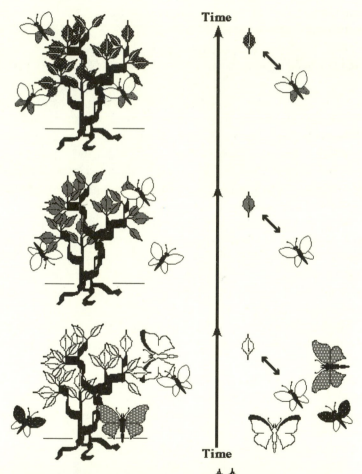

As time passes the plant evolves a succession of biochemicals [ ] that poison the butterfly larva, while the butterfly evolves enzymes [ ] that neutralize or digest these biochemicals.

Figure 1.12   A Biological Arms Race.

nonlinearity. The recycling of resources by the aggregate behavior of a diverse array of agents is much more than the sum of the individual actions. For this reason it is difficult to evolve a single agent with the aggregate's capabilities. Such complex capabilities are more easily approached step by step, using a distributed system. This is a point to be emphasized later when we examine the emergence of *default hierarchies* in the next chapter. It should be evident then that we will not find *cas* settling to a few highly adapted types that exploit all opportunities. Perpetual novelty is the hallmark of *cas*.

INTERNAL MODELS (*MECHANISM*)

In introducing mimicry, I mentioned the role of learned avoidance in birds: insectivorous birds anticipate the bitter taste of butterflies with a particular orange and black wing pattern. Just how do they do this? This question, enlarged to encompass all *cas*, takes us to another hallmark of *cas*: they anticipate. To understand anticipation we have to understand a mechanism that is itself complex—an internal model. I use *internal model* to cover much the same ground that Gell-Mann (1994) covers with his *schema*. Unfortunately, the word "schema" has become a fixture in the study of genetic algorithms, designating a related but different topic. Since both topics appear in this book, I choose to avoid confusion by using the term "internal model" to refer to the mechanism for anticipation.

The use of models for anticipation and prediction is a topic that, in its broadest sense, encompasses much of science. It is a difficult topic, but not impenetrable. In the next chapter we will bring out sufficient apparatus to discuss the generation of models, but there are some simpler aspects that we can look at now.

The basic maneuver for constructing models was pointed up in our earlier examination of aggregation: eliminate details so that selected patterns are emphasized. Because the models of interest here are interior to the agent, the agent must select patterns in the torrent of input it receives and then must convert those patterns into changes in its internal structure. Finally, the changes in structure, the model, must enable the agent to anticipate the consequences that follow when that pattern (or one like it) is again encountered. How can an agent distill experience into an internal model? How does an agent unfold the model's temporal consequences to anticipate future events?

To make a start on these questions, let's take a closer look at models as predictors. We usually ascribe prediction only to "higher" mammals, rather than taking it as a property of all organisms. Still, a bacterium moves in the direction of a chemical gradient, implicitly predicting that food lies in that direction (see Figure 1.13). The mimic survives because it implicitly forecasts that a certain pattern discourages predators. When

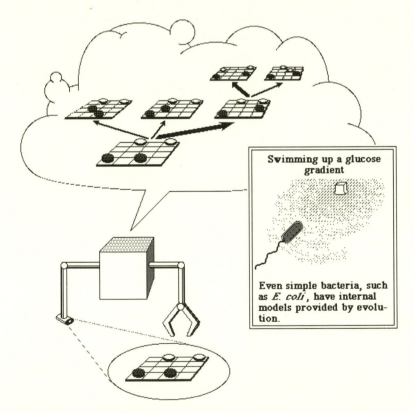

Swimming up a glucose gradient

Even simple bacteria, such as *E. coli*, have internal models provided by evolution.

Figure 1.13   Internal Models.

we get to the so-called higher mammals, the models do depend more directly on the agent's sensory experience. A wolf bases its movements on anticipations generated by a mental map that incorporates landmarks and scents. Early humans built Stonehenge as an explicit, external model that helped predict the equinoxes. Now we use computer simulations to make predictions ranging from the flight characteristics of untried aircraft to the future gross domestic product. In all these cases prediction is involved, and in the last two cases external models augment internal models.

Taking these examples into account, we will find it useful to distinguish two kinds of internal models, *tacit* and *overt*. A tacit internal model simply prescribes a current action, under an *implicit* prediction of some desired future state, as in the case of the bacterium. An overt internal model is used as a basis for *explicit*, but internal, explorations of alternatives, a process often called *lookahead*. The quintessential example of lookahead is the mental exploration of possible move sequences in chess prior to moving a piece. Both tacit and overt models are found in *cas* of all kinds—the actions and identity supplied by an immune system fall at the tacit end of the scale, whereas the internal models of agents in an economy are both tacit and overt.

How do we distinguish an internal model from other pieces of internal structure that have nothing to do with modeling? We might start with the critical characteristic of a model: a model allows us to infer something about the thing modeled. Following this line, we could say that a given structure in an agent is an internal model if we can infer something of the agent's environment merely by inspecting that structure. Certainly we can infer a great deal about the environment of any organism by studying relevant pieces of morphology and biochemistry. Accordingly, we might say that those pieces constitute a tacit internal model. But, equally, we can infer a meteorite's history from its composition and surface condition. It is clearly fruitless, even metaphorically, to attribute an internal model to a meteorite, so we need something more in our definition.

There is an additional requirement that will eliminate meteorites and other inanimate structures. We can require that the structure from which we infer the agent's environment also actively determine the agent's behavior. Then, if the resulting actions anticipate useful future consequences, the agent has an effective internal model; otherwise it has an ineffective one. With an appropriate way of connecting future credit to current actions, evolution can favor effective internal models and eliminate ineffective ones.

Despite the apparent and real differences between the bacterium's tacit model and mammalian overt models, there are important commonalities. In both cases the organism's chances of survival are enhanced by the predictions, implicit or explicit, that the model entails. Thus, variants of the model are subject to selection and progressive adaptation. The timescale for change of the implicit model of the bacterium or the mimic is orders of magnitude different from the timescale for change of a mammal's central nervous system, but the *process* of selective emphasis that generates these models is not so different as we shall see.

BUILDING BLOCKS (*MECHANISM*)

In realistic situations an internal model must be based on limited samples of a perpetually novel environment. Yet the model can only be useful if there is some kind of repetition of the situations modeled. How can we resolve this paradox?

We get the beginnings of an answer when we look to a commonplace human ability, the ability to decompose a complex scene into parts. When we do this, the component parts are far from arbitrary. They can be used and reused in a great variety of combinations (see Figure 1.14), like a child's set of building blocks. Indeed, it is evident that we parse a complex scene by searching for elements already tested for reusability by natural selection and learning.

Because reusability means repetition, we begin to see how we can have repetition while being confronted with perpetually novel scenes. We gain experience through repeated use of the building blocks, even though they may never twice appear in exactly the same combination. By way of example, consider the common building blocks for a human face: hair, forehead, eyebrows, eyes, and so on (see Figure

7 Building Blocks
(multiple copies of each)

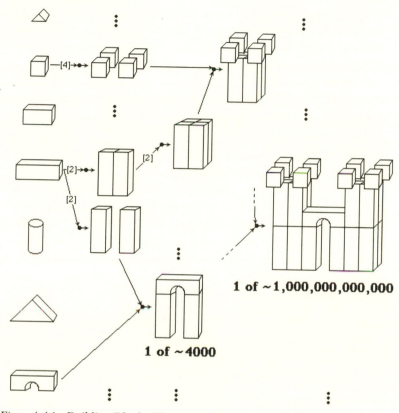

1 of ~1,000,000,000,000

1 of ~4000

Figure 1.14   Building Blocks (Generators).

1.15). Let's decompose the face into ten components (one of which is "eyes"), and let's allow ten alternatives for each component (as in "blue eyes," "brown eyes," "hazel eyes," . . .). We can think of ten "bags" holding ten building blocks each, for a total of $10 \times 10 = 100$ building blocks. Then we can construct a face by choosing one building block from each bag. Because there are ten alternatives in each bag, we can construct any of $10^{10} = 10$ billion distinct faces with these 100 building blocks! Almost any new face we encounter can be

closely described by an appropriate choice from the set of 100 building blocks.

If model making, broadly interpreted, encompasses most of scientific
activity, then the search for building blocks becomes *the* technique for
advancing that activity. At a fundamental level, we have the quarks of
Gell-Mann (1994). Quarks can be combined to yield nucleons, the
building blocks at the next level. The process can be iterated, deriving
the building blocks at successive levels from specific combinations of
building blocks at the previous level. The result is the quark /
nucleon / atom / molecule / organelle / cell / . . .   progression that
underpins much of physical science.

We gain a significant advantage when we can reduce the building
blocks at one level to interactions and combinations of building blocks
at a lower level: the laws at the higher level derive from the laws of the
lower-level building blocks. This *does not* mean that the higher-level
laws are easy to discover, any more than it is easy to discover theorems in
geometry because one knows the axioms. It *does* add a tremendous
interlocking strength to the scientific structure. We'll come back to this
point when we discuss *emergence* in *cas*.

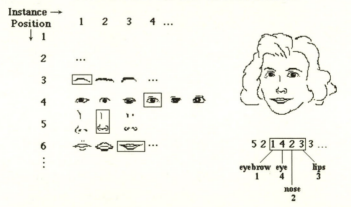

BUILDING BLOCKS AND RECOMBINATION

A face can be described by stringing together the numbers that index
its component parts.

Figure 1.15   Building Blocks for Faces.

It would be a mistake to confine our attention to the building blocks of physics. Wherever we turn, building blocks serve to impose regularity on a complex world. We need only look at the use of musical notation to transmit the endless variety of music, or the use of a limited range of morphologies to describe the tremendous spectrum of animal structures. The point applies with at least as much force to our everyday encounters. If I encounter "a flat tire while driving a red Saab on the expressway," I immediately come up with a set of plausible actions even though I have never encountered this situation before. I cannot have a prepared list of rules for all possible situations, for the same reason that the immune system cannot keep a list of all possible invaders. So I decompose the situation, evoking rules that deal with "expressways," "cars," "tires," and so on, from my repertoire of everyday building blocks. By now each of these building blocks has been practiced and refined in dozens or hundreds of situations. When a new situation is encountered, I combine relevant, tested building blocks to model the situation in a way that suggests appropriate actions and consequences.

This use of building blocks to generate internal models is a pervasive feature of complex adaptive systems. When the model is tacit, the process of discovering and combining the building blocks usually proceeds on an evolutionary timescale; when the model is overt, the timescale may be orders of magnitude shorter. Still, to reemphasize the point made both for internal models and in the initial discussion of adaptation, the underlying adaptive process remains much the same throughout the range of *cas*.

Where Next?

The next three chapters combine these seven basics (see Figure 1.16) in different ways to achieve two goals. The first goal, the object of the next chapter, is to provide a definition of "adaptive agent" that works for all the different kinds of agents found in *cas*. The second goal, to be pursued in Chapters 3 and 4, is to provide a computer-based model that has enough generality to allow us to carry out thought experiments relevant to all *cas*. We'll see that the seven basics appear over and over again, suggesting mechanisms and directions (see Figure 1.17).

Beyond these two goals is a larger objective: to uncover general principles that will enable us to synthesize complex *cas* behaviors from simple laws. Complex adaptive systems are quite different from most systems that have been studied scientifically. They exhibit coherence under change, via conditional action and anticipation, and they do so

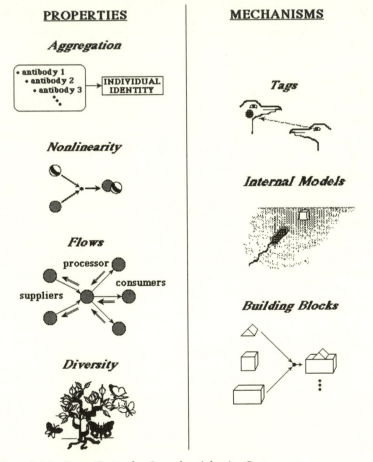

Figure 1.16   Seven Basics for Complex Adaptive Systems.

without central direction. At the same time, it would appear that *cas* have lever points, wherein small amounts of input produce large, directed changes. It should be easier to discover these lever points if we can uncover general principles that govern *cas* dynamics. Knowing more about lever points would, in turn, provide us with guidelines for effective approaches to *cas*-based problems such as immune diseases,

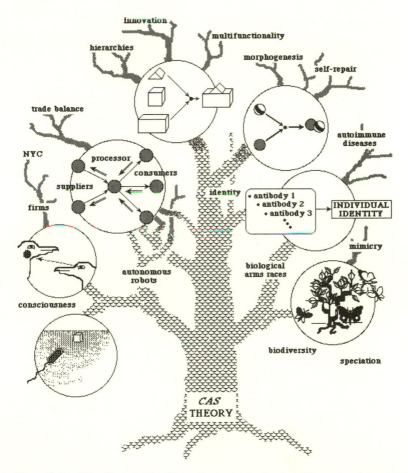

Figure 1.17   The Role of the Seven Basics in the Study of Complex Adaptive Systems.

inner-city decay, industrial innovation, and the like. For problems so complex, it is unlikely that we will make substantial progress without theoretical guidelines. We are only at the beginning of the search for general principles, but we do have some hints as to what those principles might be. I'll set down those hints, as I see them, in the concluding chapter.

· 2 ·

Adaptive Agents

W<small>E RETURN</small> to New York City for a quick illustration of the outlook provided by the seven basics of the previous chapter. Agents formed by *aggregation* are a central feature, typified by firms that range from Citibank and the New York Stock Exchange to the corner deli and the yellow cab. These agents determine virtually every fiscal transaction, so that at one level of abstraction the complex adaptive system that is New York City is well described by the evolving interactions of these agents. We have only to look to advertising, trademarks, and corporate logos to see how *tags* facilitate and direct these transactions. The *diversity* of these tags underscores the variety in the city's firms and activities, and the complex *flow* of goods into, out of, and through the city that results. That New York retains both a short-term and a long-term coherence, despite diversity, change, and lack of central direction, is typical of the enigmas posed by *cas*. As is usual, *nonlinearities* lie near the center of the enigma. New York's nonlinearities are particularly embodied in the *internal models*—models internal to the firms—that drive transactions. These models range from spreadsheets to sophisticated corporate plans. There are also continual innovations, such as the steady flux of new financial instruments on Wall Street ("derivatives," the current innovation, have absorbed even more money than their predecessors, "junk

41

bonds"). Trend projection and other linear analyses provide few insights into these activities. New perceptions will surface, I suspect, if we can uncover the *building blocks* that are combined and recombined to determine the city's outward appearance. The building blocks for this enterprise are less obvious than for some other *cas*, though contracts, organization charts, permissions, pieces of city infrastructure, and taxes are all obvious candidates.

This view of New York City is no less intricate than other ways of describing this urban setting, but it does suggest that the city is not all that different from other *cas*. We have already seen these same basic characteristics in various *cas*, and it is not particularly difficult to locate them in still others. They are distinctive, and I know of no systems that are not complex adaptive systems in which all seven are present simultaneously. That does suggest treating all *cas* within a common framework that exploits these basics. However, there is one feature of *cas* that tempers this suggestion. The agents in different systems, even within the same system, exhibit real dissimilarities. Firms in a city don't seem to have much in common with antibodies, and organisms in an ecosystem don't look at all like neurons in the nervous system. Is it really possible to find a common representation for these very different agents? If so, a uniform approach to *cas* is feasible; if not, a uniform approach seems unlikely. A common representation for agents, then, is our next objective.

Let's explore the possibilities in three stages. First, we'll look for a uniform way to represent the capabilities of different kinds of agents, without any concern for changes that might be produced by adaptation. I'll call the result a *performance system*. The next stage is to use the agent's successes (or failures) to assign credit (or blame) to parts of the performance system. I'll call this process *credit assignment*, following usage in other studies of learning and adaptation. The last stage concerns making changes in the agent's capabilities, replacing parts assigned low credit with new options. For reasons that will become apparent, I'll call this procedure *rule discovery*.

A Performance System

The first step in arriving at a common description of agents is actually a return to the description of adaptive agents in the early part of the last chapter. There we used *rules* as a descriptive device; now we take rules more seriously as a formal means of defining agents. For the rules to be a successful unifying device, applicable whatever the agent's outward form, they must meet three criteria:

1. The rules must use a single syntax to describe all *cas* agents.
2. The rule syntax must provide for all interactions among agents.
3. There must be an acceptable procedure for adaptively modifying the rules.

As in the last chapter, we look first at the simplest kind of rule: IF (some condition is true) THEN (execute some action). IF/THEN rules are used for explanation in a wide variety of fields: in psychology they are called stimulus-response rules (see Figure 2.1); in artificial intelligence they are called condition–action rules; and in logic they are called production rules. Our immediate objective is to find a simple

STIMULUS - RESPONSE

IF   **SMALL FLYING OBJECT TO LEFT**

THEN   **TURN HEAD 15° LEFT**

Figure 2.1   A Stimulus-Response Rule.

syntax for IF/THEN rules, a syntax that will work for any kind of agent. Later we will add a simple modification that gives IF/THEN rules enough power to model any agent that can be modeled on a computer.

Input/Output

The syntax we use for the IF/THEN rules depends critically on the way an agent interacts with its environment. Let's start with the input side. In ordinary terms, an agent senses the environment via an assortment of stimuli. If the agent is an antibody, the stimuli are the molecular configurations—tags—on the surfaces of the antigens. If the agent is a human, the stimuli come through the five senses. If the agent is a business firm, the stimuli are orders, cash flow, incoming goods, and so on. Typically, an agent is inundated with stimuli, receiving far more information than can be put to use.

The agent's first task, then, is to filter the torrent of information its environment produces. To describe this filtering operation, I adopt the common view that the environment conveys information to the agent via a set of *detectors*. The simplest kind of detector is one that senses a particular property in the environment, turning "on" when the property is present and "off" when it is not (see Figure 2.2). That is, the detector is a binary device that conveys one bit of information about the environment. Such detectors might seem quite limited in their ability to sense the environment, but an arbitrarily large amount of information can be conveyed by a sufficiently large cluster of detectors. Indeed, the amount of information conveyed goes up exponentially with the number of detectors. A set of three binary detectors can code for $2 \times 2 \times 2 = 2^3 = 8$ colors; a set of twenty detectors, using a variant of the "Twenty Questions" game, can produce a unique stimulus for each of 2^{20}, more than a million, distinct categories.

It is worthwhile to emphasize, concerning detectors, a caution earlier invoked for rules. This discussion of detectors is *not* a claim that all *cas* agents *use* binary detectors. It is, rather, a claim that we can use clusters of binary detectors to *describe* the way agents filter information from the environment; we can translate other means of detection into this format. The value of binary detectors in this discussion rests on their usefulness in modeling arbitrary adaptive agents.

By means of binary detectors we can use standardized messages, binary strings, to represent the information selected by the agent. Can we extend this standardization to the agent's output side? The actions of *cas* agents are, after all, as various as their ways of extracting information from the environment. We can gain some ground in regularizing output by "inverting" the function performed by detectors. Let me describe the agent's actions in terms of a set of *effectors*. Each effector has an elementary effect on the environment when it is activated by an appropriate message (Figure 2.2). At any given time, the overall response of the agent is generated by the cluster of effectors active at that time. That is, the effectors *decode* standardized messages to cause actions in the environment. In so doing, the effectors "invert" the procedure used by the detectors to *encode* environmental activity into standardized messages. As with detectors, we use effectors as a descriptive device for modeling the adaptive agent output.

PROCESSING AND SYNTAX

With this description of the input and output of agents in terms of messages, it seems advantageous to handle interactions of the agent's rules in the same way. Providing for rule interaction is the critical step

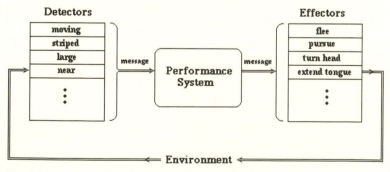

Figure 2.2   Detectors and Effectors for a Performance System.

that gives simple IF/THEN rules the full power of a programming language. For one IF/THEN rule to interact with another, it must be that the IF part of one of the rules is sensitive to the actions specified by the THEN part of the other rule. If we think of each rule as a kind of microagent, we can extend the input/output role of messages to provide for interactions. Think of each rule as having its own detectors and effectors or, more to the point, think of each rule as a message-processing device. The rule then has the form

> IF (there is a message of the right kind) THEN (send a specified message).

That is, an agent is described now as a collection of *message-processing rules* (see Figure 2.3). Some rules act on the detector-originated messages, processing information from the environment, and some rules act on messages sent by other rules. Some rules send messages that act on the environment, through the agent's effectors, and some rules send messages that activate other rules (see Figure 2.4).

IF   SMALL FLYING OBJECT CENTERED
THEN   SEND @

IF   @
THEN   EXTEND TONGUE

A message, represented here by the uninterpreted symbol @, is typically represented by an uninterpreted bit string in implementations.

Figure 2.3   A Small Message-Passing Rule-Based System.

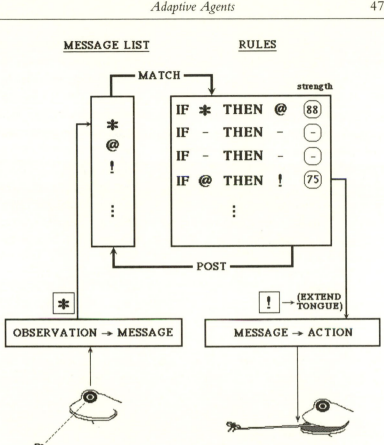

Figure 2.4   A Message-Passing Performance System.

With this description as a guide, we can develop a general syntax for *cas* agents (see Figure 2.5). We begin with the allowable messages. For simplicity of exposition, assume that all messages are binary strings, strings of 1's and 0's, and that they are all of standard length. (The last assumption means that messages are much like the binary strings stored in the registers of a computer.) Neither of these assumptions is really necessary, but neither causes any serious loss of generality—and they do simplify presentation. Notationally, a message has the form

10010111 . . . 1

$$| \; \leftarrow \; L \; \rightarrow \; |$$

where L is the length of the standard message. The set of all possible messages, M, is thus the set of all strings of 1's and 0's of length L. The formal designation of this set is $\{1,0\}^L$.

Next we must provide a syntax for the condition side of the rules, a syntax that specifies which messages the rule responds to. Again, there are many ways to do this, but one of the simplest is to introduce a new symbol #, which can be interpreted as "anything is acceptable at this position." More colloquially, it is a "don't care" symbol. Consider the string of symbols

1#### . . . #

$$| \; \leftarrow \; L \; \rightarrow \; |$$

used as the condition part of a rule. This condition responds to any message that starts with a 1, not caring what digits appear at the other $L-1$ positions. Similarly, the string

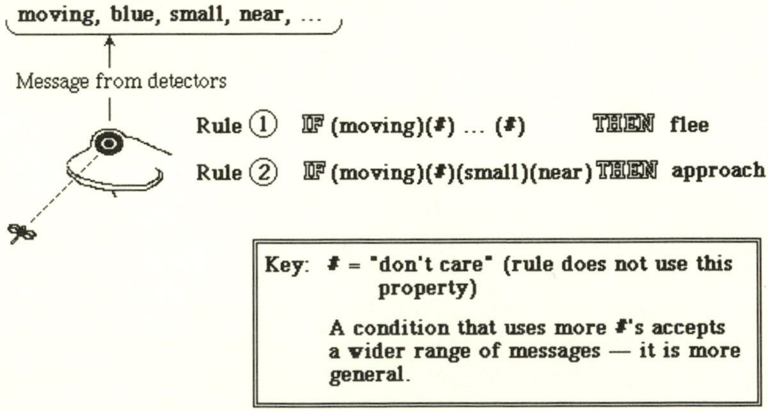

Figure 2.5   Syntax for a Performance System.

$\#\#\#\# \ldots \#\#1\#0$

$|\quad\leftarrow\quad L\quad\rightarrow\quad|$

represents a condition that responds to any message that has a 1 at the second to last position, $L-2$, and a 0 at the last position, L. Under this arrangement the set of all possible conditions, C, is the set of all strings of 1's, 0's, and #'s of length L. The formal designation of this set is $\{1,0,\#\}^L$.

Because the only action of a rule in this format is to post a message, all rules have the form

IF (condition from C satisfied) THEN (send message from M).

For example, with $L=5$, the rule

IF (1####) THEN (00000)

will transmit the message 00000 if it detects any message starting with a 1. The similar rule

IF (10101) THEN (00000)

will transmit the message 00000 only if it detects the specific message 10101.

With the two sets, $M = \{1,0\}^L$ and $C = \{1,0,\#\}^L$, and this format for rules, we have the capacity to describe the behavior of a wide variety of agents. A particular agent is described by setting down the cluster of rules, in this fixed format, that generates its behavior. Rules so defined act much as instructions in a computer, the cluster serving as a program that determines the agent's behavior. If there is any way to model an agent on a computer, these technical conditions guarantee that it can be modeled using a cluster of rules in this format. To get full computational power we must give our rules two independent conditions, IF ( ) AND IF ( ) THEN ( ), and provide them with negation, IF

NOT (   ) THEN (   ), but we can ignore these refinements for present purposes.

With this syntax we have a uniform, rule-based technique for modeling agents, be they neurons, antibodies, organisms, or business firms. Figures 2.1 and 2.3, not to be taken too seriously, illustrate the use of a rule or two to capture one facet of the behavior of a frog (the abstract symbols emphasize the arbitrariness of the bit-strings that encode the messages).

SIMULTANEOUS ACTIVITY—PARALLELISM

Before proceeding further we must make a careful distinction between the different uses of messages in this system. The detector-originated messages have a built-in meaning assigned by the environmental properties detected. The rule-originated messages, on the other hand, have no assigned meaning except when they are used to activate effectors. They acquire meaning in terms of their ability to activate other rules. It is important to differentiate these two kinds of messages. Otherwise, rule-originated messages might be taken as coming from the environment, producing "hallucinations" for the agent. The distinction is usually accomplished by assigning identifying tags to the two kinds of messages.

Because rule-originated messages have no built-in meanings (setting aside for the moment messages that activate effectors), we are not faced with contradictions when several rule-originated messages are present at the same time. That means we can have several rules active simultaneously without fear of contradiction; more rules active simply mean more messages. This is a substantial advantage. We have a system that can model the concurrent activities typical of *cas* and, as we will see, we can use messages as building blocks for modeling complex situations.

To exploit this advantage we provide the agent with a kind of inventory, a *message list*, that stores all current messages. A useful, if somewhat fanciful, metaphor for thinking about an agent's performance under this arrangement is an office in which there is a large bulletin board. The workers in the office are assigned desks, each of which has responsibility for responding to certain kinds of memos on the bulletin board. And, of course, the output of each desk is more memos. At the beginning of the day the workers take down the memos, they process them throughout the day, and at the end of the day they post the new memos that have resulted from their efforts. In addition, some memos come in from outside the office, and some memos go from the office to the outside. Under this metaphor, the agent corresponds to the office, the memos to messages, the bulletin board to the message list, the desks to rules, memos from outside the office to detector-originated messages, and memos to the outside correspond to effector-directed messages. In the agent, as in the office, many activities go on simultaneously, and only some of them are visible on the outside.

This provision for simultaneously active rules helps us understand an agent's ability to handle a perpetually novel world. It contrasts sharply with an approach wherein the agent has only a single rule for each situation. With the single-rule approach, the agent must have rules prepared for every situation it may plausibly encounter. This poses a problem analogous to the one we discussed earlier for the immune system. An agent is unlikely to have a single rule adequate for each situation it encounters for the same reason that the immune system lacks a set of antibodies prepared ab initio for all possible invading antigens—there are just too many possibilities. With simultaneously active rules, the agent can combine tested rules to describe a novel situation. The rules become building blocks.

By way of example, consider someone in the unfortunate circumstance of having a "flat tire while driving a red Saab on the expressway." Most of us have not driven a Saab, let alone had a flat tire while driving one, but we would not be at a loss for an appropriate response. The reason would seem to be that we decompose the situation into familiar parts. Most of us have had some experience with flat tires, or at least know procedures for dealing with them. Most of us have driven on an expressway. And so on. We can describe this in terms of rules for dealing with components of the situation. In terms of our syntax for rule-based agents, this means rules of the form IF (flat tire while driving) THEN (slow down), IF (on an expressway with a flat) THEN (pull into breakdown lane), and so on, encoded in the *C / M* syntax (see Figure 2.6). These rules, evoked simultaneously by the detector-originated messages and by other rules, activate the appropriate effector sequences. Of course, in a real situation there would be many overtones not captured in this simple example. There would be messages and active rules corresponding to short-term memory (recent happenings on the expressway), objectives of the trip, and so on. Hundreds of rules might be active, but the principle of decomposing the situation, and relevant history, into familiar building blocks would be the same.

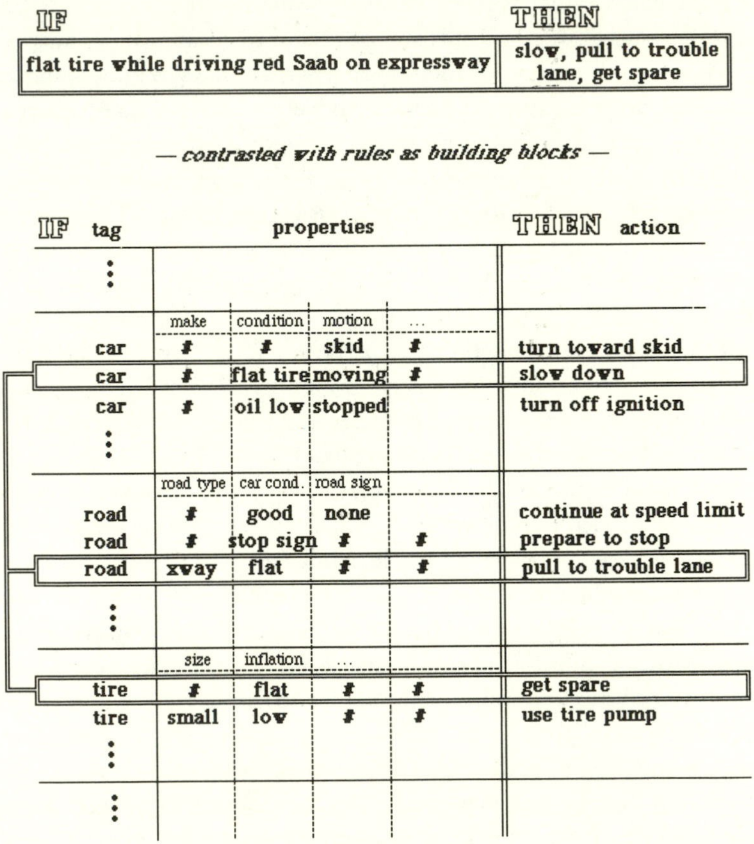

IF	THEN
flat tire while driving red Saab on expressway	slow, pull to trouble lane, get spare

— contrasted with rules as building blocks —

IF tag   properties   **THEN** action

tag	make	condition	motion	...	action
⋮					
car	*f*	*f*	skid	*f*	turn toward skid
car	*f*	flat tire	moving	*f*	slow down
car	*f*	oil low	stopped		turn off ignition
⋮					

tag	road type	car cond.	road sign		action
road	*f*	good	none		continue at speed limit
road	*f*	stop sign	*f*	*f*	prepare to stop
road	xway	flat	*f*	*f*	pull to trouble lane
⋮					

tag	size	inflation	...		action
tire	*f*	flat	*f*	*f*	get spare
tire	small	low	*f*	*f*	use tire pump
⋮					

Figure 2.6   An Example of Rule Parallelism.

Adaptation—By Credit Assignment

We have said nothing so far about the agent's ability to adapt. We have described the agent's *performance system*, its capabilities at a particular point in time. Now we have to look into ways of changing the system's performance as it gains experience.

The first step is to look more closely at the role of rules in the performance system. The usual view is that the rules amount to a set of facts about the agent's environment. Accordingly, all rules must be kept consistent with one another. If a change is made or a new rule is introduced, it must be checked for consistency with all the other rules.

There is another way to consider the rules. They can viewed as hypotheses that are undergoing testing and confirmation. On this view, the object is to provide contradictions rather than to avoid them. That is, the rules amount to alternative, competing hypotheses. When one hypothesis fails, competing rules are waiting in the wings to be tried. My inclination is toward this latter view.

If there is to be a competition, there must be some basis for resolving it. It is also clear that the competition should be experience based. That is, a rule's ability to win a competition should be based on its usefulness in the past. The objective is closely related to the statistician's concept of building confirmation for a hypothesis. We want to assign each rule a *strength* that, over time, comes to reflect the rule's usefulness to the system. The procedure for modifying strength on the basis of experience is often called *credit assignment*.

Credit assignment is a relatively easy task when the environment produces direct payoff (reward, reinforcement) for an action. If we turn a key and the car starts, that action quickly becomes part of our repertoire. Credit assignment is much more difficult when some early stage-setting action makes possible a later useful outcome. The problem is clearly exposed if we examine the play of a board game, say checkers. Taking a triple jump in checkers, when possible, almost always leads to a win and, as with the ignition key, it is easy to credit a rule that takes that action. But how should a neophyte credit a rule when that rule's action is followed *four moves later* by the triple jump option? How does the neophyte know it was *that* rule, not some rule acting earlier or later, that was critical in setting the stage? Or perhaps the outcome was simply a mistaken move on the part of the opponent. Yet good play in checkers, and sophisticated actions in *cas*, depend on crediting anticipation and stage setting.

The credit-assignment problem becomes still more complicated when we consider a performance system with many rules active simultaneously. As the system continues to adapt, some rules will be useful and some will not. Some will decompose the environment in ways that offer useful guides to action and some will not. Moreover, long periods often elapse before the consequences of current action are obvious. Some actions can be hurtful in the short run but helpful in the long run, much like a gambit in chess. With all of these impediments, how does an agent determine which rules are helpful and which are obstructive?

Here we can use another metaphor to advantage, a standard link between competition and capitalism. Each rule can be treated as a producer (factor, middleman) buying and selling messages. The "suppliers" to a rule are those that send messages satisfying its condition(s); the "consumers" for a rule are those that act on its message. A rule's strength is treated as its cash in hand. When a rule buys a message, it must pay for it from its cash in hand; that is, its strength is reduced. When a rule sells a message, its strength is increased by the amount paid to it by the buyer (see Figure 2.7).

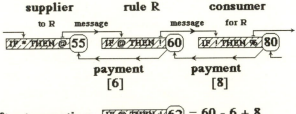

Stage-setting rules leading to reward become strong.

Figure 2.7   Credit Assignment—Changing Rule Strength.

Competition is introduced through a bidding process (see Figure 2.8). Only rules that have their conditions satisfied are eligible to bid, and only the winners gain the right to post ("sell") their messages. The size of a rule's bid is determined by its strength. Stronger rules bid more. The winners then pay their suppliers; the losers pay nothing.

After winning, the winners have less strength and their suppliers have more strength. The winners, however, have gained the right to post their messages, with the possibility that they will have consumers that will bid and pay. In this setting, a winning rule will thrive—get stronger—only if its consumers pay more than the amount bid in the first place. The old capitalist adage holds: buy cheap and sell dear!

Just how does this spate of buying and selling help the adaptive agent solve its credit-assignment problem? To make the connection, we must determine the ultimate consumers (buyers). They are the rules that are active when the agent receives an overt reward from the environment. The agent knows that these actions are desirable, as in the case of the triple jump, so the rules directly responsible are automatically strengthened. We can think of the overt reward as being shared among the rules

IF OBJECT TO LEFT **THEN** TURN HEAD 15° *LEFT* $\boxed{88}$

IF OBJECT TO LEFT **THEN** TURN HEAD 15° *RIGHT* $\boxed{12}$

Rules act as competing hypotheses; the stronger the rule the more likely it is to win the competition.

Only winning rules post their messages.

Figure 2.8 Rule Competition in a Parallel Rule-Based System.

active at the time of reward. This is much like Pavlovian conditioning, with immediate reinforcement of desirable actions.

Now consider any rule that is an immediate supplier of a strengthened "ultimate-consumer" rule. Assume that this supplier helps set the stage, making it possible for the ultimate-consumer rule to evoke a reward from the environment. As the rewards make the ultimate-consumer rule stronger, it makes larger bids because its bids are proportional to its strength. The supplier in turn becomes stronger because of the larger payments it receives. After a while, the suppliers of the supplier will benefit from this increasing strength if they set the stage for the supplier. We can iterate this argument over any chain of suppliers that progressively sets the stage for some overtly rewarding action. All rules in the chain will eventually be strengthened because of the progressive strengthening of their consumers.

A question: What if the supplier rule sends a message that activates an ultimate-consumer rule, but "cheats" by not appropriately setting the stage for the consumer's action? The consumer rule will then, of course, not be rewarded, even though it has paid its supplier. It will have paid without being paid, with a corresponding reduction in its strength. As a consequence, the next time around, the cheating supplier will be paid less by the consumer. Because the supplier is earlier in its strength-building process than the ultimate-consumer rule, its strength will soon fall below the point where it can win competitions. This is particularly true if there are other rules that *do* set the stage for the ultimate-consumer rule. Cheaters do not thrive under this regime. Again, this argument can be iterated over any chain of suppliers.

This credit-assignment procedure, which I call a *bucket brigade algorithm*, strengthens rules that belong to chains of action terminating in rewards. The process amounts to a progressive confirmation of hypotheses concerned with stage setting and subgoals. Theorems from mathematical economics can prove this outcome for statistically regular environments, and computer simulations show that it works in a wide variety of environments, particularly when combined with the rule discovery process.

INTERNAL MODELS

There is a modification to the bidding process that furthers the construction of internal models. It is based on the intuition that, other things being equal, an agent should prefer rules that use more information about a situation. In our syntax, the amount of information used by a rule depends upon the number of #'s in the rule's conditions. A rule is more *specific* if it has fewer #'s in its conditions (see Figure 2.5). For instance, the condition ## ... # accepts any message, so it provides no information whatsoever when it is satisfied. At the other extreme, the condition 11 ... 1 is satisfied by one specific message, a string of 1's, providing the maximum possible information. To implement the preference we must modify the bidding process. The simplest way is to make the bid proportional to the *product* of strength and specificity. That way, if *either* the strength *or* the specificity is close to zero, the bid will be close to zero; only if *both* are large will the bid be large.

Consider now a competition between a more specific rule, $r1$, and a less specific rule, $r2$. For a concrete example (see Figure 2.9), let $r1$ be the stimulus–response rule

IF (there's a moving object in the environment) THEN (flee),

and let $r2$ be the stimulus–response rule

IF (there's a small moving object nearby in the environment) THEN (approach).

Any message concerning a moving object will satisfy $r1$, but only a subset of those messages will satisfy $r2$, namely those messages proclaiming the additional properties that the object is small and nearby. However, when there is a small moving object nearby, $r1$ and $r2$ will be in direct competition. If $r1$ and $r2$ are roughly equal in strength, $r2$ will have the advantage because of its higher specificity. That is, $r2$ makes a bigger bid because it uses more information about the situation.

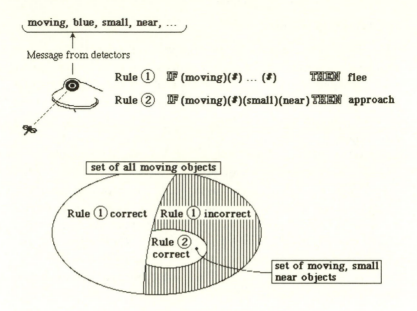

Figure 2.9   A Rule-Based Default Hierarchy.

We are now in a position, for the first time, to discuss the formation of internal models. In effect, the two rules *r*1 and *r*2 form a simple model of the environment. It is an apparently unresolved model, because *r*1 and *r*2 are contradictory when they are active simultaneously. However, a closer look at this contradiction reveals a kind of symbiosis between these two rules. Assume this agent, a "frog," lives in an environment where most moving objects, "herons" and "raccoons," are dangerous, but small moving objects, "flies," are prey. The more general rule, *r*1, becomes a kind of default to be used when detailed information is lacking: "If it's moving, it's dangerous." Still, if this rule were always invoked, the frog would starve to death because it would flee its food, flies as well as everything else. The more specific rule, *r*2, on the other hand, advocates the correct action when flies are around. It provides an exception to the default rule, and because it is more specific it outcompetes the default when the additional constraints "small and nearby" are present. The following argument reveals the symbiosis. Every time the default *r*1 makes a mistake, it loses strength. When *r*2 wins, preventing the mistake, it saves $r1$ the loss. Thus, the presence of $r2$, though it contradicts $r1$, actually benefits $r1$. The two rules together provide the frog with a much better model of the environment than either alone would provide.

In forming internal models with the present syntax, we will find it easier to discover and test a general rule than a specific one. To see this, consider an agent that has $L = 100$ detectors. The simplest condition that uses any information at all is one that relies on a single detector, having #'s for all other detectors. A case in point would be the default rule for our frog, which uses only the property "moving." Just how many distinct conditions are there that rely on only one of the 100 detectors? We can count them as follows. Select any one of the 100 detectors (positions) as the property we're interested in. We then have to decide whether the condition is to require the property to be present (1) or absent (0). That is, we can select any one of 100 positions, and there are two possibilities for the position. So there are just 200 different possible conditions that use only a single detector. All 200 of these conditions could be tested for usefulness in a short time.

At the other extreme is a condition that uses all of the detectors. Here, we have to select one of the two possibilities, present (1) or absent (0), for *each* of the 100 positions. So there are

$$2 \times 2 \times 2 \ldots \times 2 = 2^{100} \cong 10^{30}$$

$$| \quad \leftarrow \quad 100 \quad \rightarrow \quad |$$

distinct conditions of this kind. This huge number is much larger than the estimated lifetime of the universe measured in microseconds. Clearly, it is not feasible for an agent to try *all* such conditions.

General conditions are not just fewer in number, they are also tested more frequently by the agent in typical environments. As a trial, let's assume that all messages from the detectors are equally likely. Then a given detector will be on (1) about as frequently as it is off (0). This is the same as saying that about half of all messages will have a given value, say 1, for a given detector. Consider then the general condition

1#.#.#. . . . #. It will be satisfied about half the time! It is being tested quite frequently, so credit assignment will quickly designate an appropriate strength to a rule using this condition.

Contrast the only slightly more specific condition 10### . . . #. Half of the messages will contain a 1 at the first position, but only half of those will also have a 0 at the second position. That is, only $\frac{1}{2} \times \frac{1}{2} = \frac{1}{4}$ of the messages will satisfy 10### . . . #, so that condition gets tested only half as often as 1### . . . #. It is easy to see that the testing rate drops by $\frac{1}{2}$ for *each* additional detector value used by the condition.

DEFAULT HIERARCHIES

Obviously, useful general conditions—defaults—are relatively easy to find and establish. The more specific exception rules take progressively longer to find and establish. This suggests that, under credit assignment, agents early on will depend on overgeneral default rules that serve better than random actions. As experience accumulates, these internal models will be modified by adding competing, more specific exception rules. These will interact symbiotically with the default rules. The resulting model is called a *default hierarchy* (see Figure 2.9). Of course, evolution may have "wired in" some specific rules (reflexes, for instance) produced by generations of genetic selection. It may also happen that highly specific conditions develop in response to a common, salient detector-message. But neither of these cases contradicts the principle that default hierarchies expand over time from general default to specific exceptions.

Now we need to look at the mechanisms an agent can use to generate candidates for the default hierarchy.

Adaptation—By Rule Discovery

The first process that comes to mind for rule generation is to carry out a kind of random trial and error, making limited random changes in the rules already in place. This procedure may work on occasion, but it does not make much use of system experience. Truly random changes are like coin flipping: what happens next does not depend on what has happened before. To make random changes in a complicated internal model such as a default hierarchy, in the hopes of improving it, is much like making random changes in a complicated recipe. Most changes will not be for the better.

What other options are there? We'll do better if we can assure some kind of *plausibility* for the newly generated rules: they should not be obviously wrong when viewed in the light of past experience. In most cases, plausibility arises from the use of tested building blocks. If we go back to the "flat tire while driving a red Saab on the expressway," we see that plausibility resulted from using well-known building blocks to describe the new situation. If we follow this line, the idea would be to find components—building blocks—for individual rules. Then, intuition would say, a component that consistently appears in strong rules should be a likely candidate for use in new rules. With enough strong rules, and useful ways of locating components in them, we can generate a vast number of new rules without ever departing from tested components. The new rules are only plausible candidates—they may not prove out—but the process should be considerably more efficient than random trial and error. And, of course, there may be ways of discovering new rule components, opening new ranges for testing.

A brief look at the role of tested building blocks in technical innovation will help us understand the role of building blocks in the specific case of rule innovation. A scan of history shows that technical innovations almost always arise as a particular combination of well-known building blocks. Take two technological innovations that have revolutionized twentieth-century society, the internal combustion engine and the digital computer. The internal combustion engine combines Volta's sparking device, Venturi's (perfume) sprayer, a water pump's pistons, a mill's gear wheels, and so on. The first digital computers combined Geiger's particle counter, the persistence (slow fade) of cathode ray tube images, the use of wires to direct electrical currents, and so on. In both cases most of the building blocks were already in use, in different contexts, in the nineteenth century. It was the specific combination, among the great number possible, that provided the innovation. When a new building block is discovered, the result is usually a range of innovations. The transistor revolutionized devices ranging from major appliances to portable radios and computers. Even new building blocks are often derived, at least in part, by combining more elementary building blocks. Transistors were founded on knowledge of selenium rectifiers and semiconductors.

SCHEMATA

What about building blocks for rules? The most direct approach, for the rule syntax used here, exploits the values at selected positions in the rule string as potential building blocks. For instance, we can ask whether or not it is useful, on average, to start a condition with a 1 at the first position. In the earlier example of the frog, the first position corresponds to the movement detector. For the frog, the question about using a 1 at the first position as a common building block for new rules translates to a question about the importance of movement in the environment.

This approach, treating the values at individual positions as building blocks, corresponds closely to the classical approach for evaluating the effects of individual genes on a chromosome. Each gene has several alternative forms, called *alleles*. The different alleles for the human gene for eye color, for instance, produce blue eyes, brown eyes, green eyes, and so forth. Or we can look to Mendel's experiments with pea plants (nicely described in Orel, 1984)—the experiments that founded genetics. Among the genes Mendel investigated was one that controlled the surface texture of the peas. One allele produced a smooth-surfaced pea, another produced a rough surface. Genes commonly have alternative forms, and these different forms usually have distinct observable effects on the organism. The objective in genetics, as it is for rules, is to determine the effects of different alternatives at different positions.

In mathematical genetics there is a classical approach to determining these effects. It is to assume that each allele contributes something, positive or negative, to the overall fitness of the organism. The contribution is estimated by looking at the average fitness of all the individuals carrying that allele. Smooth-surfaced peas might tend to sprout more often, so the smooth-surface allele would be assigned an appropriate positive contribution. At least in principle, we could proceed through each of the genes and alleles in this way, determining the contribution of each. The overall fitness (value, strength) of any chromosome would be the sum of the contributions of its constituent building blocks, the alleles.

There are two major difficulties with the position-by-position approach. First of all, a given allele may have different effects in different environments. Blue eyes may be valuable in the far north and detrimental at equatorial latitudes. More important, alleles interact. It is rare that the effects of any gene can be isolated, as in the special cases of eye color or surface character. Particular genes affect many characteristics and the effects of different genes overlap. In short, fitness in a given environment is a nonlinear function of the alleles.

When we change focus from genetics to IF/THEN rules, the first of these difficulties is handled automatically. The conditional part of the rule—the IF—automatically selects the "environment" in which the rule will act. So the evaluation of the parts of the rule proceeds only in the environments for which it is designed. The second difficulty—nonlinearity—is not so easily disposed of, whether in genetics or rules. I am about to propose an approach that works for both.

To begin, we must allow for building blocks that use more than a single position in the string. That is, we would like to allow a building block that encompasses the first three positions, or a building block that encompasses positions 1, 3, and 7. For our frog this could be a building block that amalgamates "moving," "small," and "nearby." We need a simple way to designate such a building block. The fact that we want to look at some specific positions and ignore others suggests that we make new use of the "don't care" symbol that was helpful in the syntax for rule conditions. Let's use a new symbol, "★," so we don't confuse the two uses. If we are interested in a building block that places a 1 in the first position of a condition, we designate that building block by

$$1\star\star\star\star \ldots \star$$

$$| \leftarrow L \rightarrow |;$$

if we are interested in a building block that places a 1 at the first position, a # at the third position, and a 0 at the seventh position, we designate that building block by

$$1\star\#\star0\star\star \ldots \star$$

$$| \leftarrow L \rightarrow |.$$

A building block defined in this way is called a *schema*; the positions in the string that contain symbols other than a * are called the *defining positions* of the schema.

Note that the # plays a very different role from the *. Recall that the set of all possible conditions for rules is specified formally as $\{1,0,\#\}^L$, the set of all strings of length L using the alphabet $\{1,0,\#\}$. Each condition specifies a set of messages it will accept. We can interpret a schema in a similar way. In defining the schema, we constrain some of the positions in the condition, the defining positions, to have one of the values from $\{1,0,\#\}$, and we make no requirement on the remaining conditions, indicating this by a *. Formally, then, the set of schemata for conditions is the set of all strings of the form $\{1,0,\#,\star\}^L$. An individual schema from $\{1,0,\#,\star\}^L$ specifies the set of all conditions that use that building block, much as an individual condition from $\{1,0,\#\}^L$ specifies the set of messages it accepts.

This mathematical convention, that the condition is identified with the set of messages it accepts, while the schema is identified with the set of conditions that contain it as a building block, helps distinguish # from *. The condition 1#111 . . . 1 accepts exactly two messages, 10111 . . . 1 and 11111 . . . 1. The schema 1*111 . . . 1, on the other hand, appears in three distinct conditions, 1#111 . . . 1, 10111 . . . 1, and 11111 . . . 1. The first of these conditions accepts the two messages, 10111 . . . 1 and 11111 . . . 1, but the second condition accepts only one message, 10111 . . . 1, and the third condition also accepts only one message, 11111 . . . 1. The * helps us define different *sets* of conditions, while the # helps us define different *sets* of messages.

Crossing Over and the Fitness of Schemata

With this notion of building blocks in hand, we can discuss the generation of plausible new rules in a careful way. It turns out that the metaphor from genetics can be extended to suggest an actual procedure. The metaphor thus far is the following. The gene positions on the chromosome correspond to the positions on the string defining the rule; different alleles correspond to the different values {1,0,#} that can be placed at each position in the rule string. We can go further. Mathematical genetics commonly assigns a numerical value, called *fitness*, to each chromosome. That value indicates the ability of the corresponding organism to produce surviving offspring, as in the case of Mendel's peas. In similar fashion, the strength assigned to a rule under credit assignment measures the rule's usefulness. If "survival" is mapped to "usefulness," then fitness corresponds closely to strength. To extend the metaphor, then, let's treat strength as the counterpart of fitness.

The extension suggests a procedure. Fit organisms are successful *parents*, producing offspring that grow to be parents in turn. This analogy suggests treating strong rules as parents. Some useful ideas follow from this correspondence.

- Offspring typically coexist with the parents, usually replacing other, weaker contenders in the environment. In a rule-based system this arrangement is important, because strong rules represent knowledge won. Under competition, strong rules usually determine the agent's actions, so they are the core of the agent's internal model.

- Offspring are *not* identical to the parents, so this is a genuine discovery process. Offspring, in both genetics and rule-based systems, amount to new hypotheses to be tested against the environment. In genetics an interaction called *crossing over* causes the characteristics of the parents to appear in new combinations in the offspring. It is this recombination of *sets* of alleles that is most interesting from the point of view of rule discovery, so we will discuss it at length.

Crossing over is the mechanism that breeders exploit when they cross-breed superior plants and animals. It is a close-to-literal description of what happens to a pair of chromosomes when they exchange genetic material. During the phase when the germ cells are being formed (meiosis), a chromosome from one parent may cross over the chromosome from the other parent, forming a kind of X-shape (this arrangement can actually be seen in micrographs of the DNA). Then, say, the "upper arms" of the X are exchanged (see Figure 2.10). The result, after separation, is a pair of chromosomes that differ from the parental chromosomes. Each contains a segment, from the "tip" to the point of crossing, from one parent and then continues to the other end with a segment from the other parent.

We know that crossing works well in combining superior characteristics of corn or race horses, but is it subtle enough to work with rules? In the case of the corn or race horses, we know what characteristics we want to enhance, and we select the parents accordingly. When we look to rule-based agents, we have no a priori list of characteristics. Our only measure is the overall strength of each rule. Individual building blocks (sets of alleles) within the rule do *not* have individual values. How can we make judgments about individual building blocks? More to the point, can crossing over implement such judgments automatically?

Let's start with the question of estimating the value of building blocks (schemata) when our only data are the strengths of whole rules. Note first that simple schemata—schemata where almost all positions are occupied by ★'s—will have many occurrences in an agent with many rules. For example, if the agent has many rules, a large portion of them will usually start with a 1. All are exemplars of the schema 1★★★ . . . ★.

Intuition would say that that schema is a useful building block if the rules that contain it are, on average, stronger than other rules. To capture this intuition precisely, we must be able to compare the average strength of the rules carrying 1★★★ . . . ★ to the overall average strength of the agent's rules. Call the average strength of all the agent's rules A. First determine A, then determine the average strength of the rules using

Crossover Operator

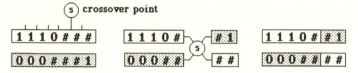

Genetic Algorithm

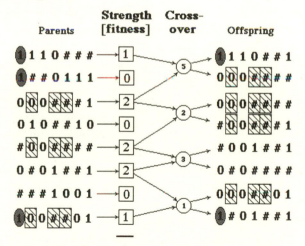

Ave. Strength of All Indivs. = 1

Ave. Strength of Instances of ●★★★★★★ = (1+0+1)/3 = 2/3

Ave. Strength of Instances of ★[0]★[##]★★ = (2+2+1)/3 = 5/3

Figure 2.10   Crossover and Genetic Algorithms.

1★★★ . . . ★. Call the latter $S(1$★★★ . . . ★$)$. We consider the schema 1★★★ . . . ★ as better than average if $S(1$★★★ . . . ★$)$ is greater than A.

Because this is only an estimation procedure, it can be wrong in particular cases. It may be that the agent's rules are peculiar in some way. For example, the agent's past experiences may not give a reliable cross-section of its environment vis-à-vis the schema 1★★★ . . . ★. Then the strengths of the rules using that schema will be skewed in some way. Human agents often operate under such misapprehensions. Nevertheless, the estimate does provide a guideline where we had none before. And if it is wrong, subsequent estimates will tend to correct the error. The procedure is much like the confirmation of a hypothesis through continued experimentation.

If we greatly simplify the relations between schemata, we can think of them as forming a kind of fantastic "landscape." Each schema is a point in the landscape, and the corresponding schema average is the height of the landscape at that point. Our objective is to find "hills" in this landscape that are higher than ones already explored. Actually, schemata as subsets of the space of possibilities form a complicated lattice of inclusions and intersections, but the landscape metaphor is a useful starting point.

Stuart Kauffman and his colleagues have studied simple versions of these landscapes—the *n-k* landscapes (see Kauffman, 1994). *N-k* landscapes have built-in statistical symmetries that make mathematical analysis possible. Analysis of these special cases, though it is not easy, does reveal some interesting guidelines, which may be generalizable to the more intricate relations that hold in the space of schemata—but that is yet to be established.

Even if the landscape metaphor can be exploited, there is still a problem. For each schema x of interest, we have to calculate the average $S(x)$ if we are to be able to estimate the value of that schema. Just how many schemata are there? The number is very large, which helps by providing many alternatives, but hinders by requiring the calculation of many averages. To get some feeling for how large the number is, let's look at the different schemata that can be found in a single condition of length L. Consider the condition

10#10# . . . 10#

$$| \ \leftarrow \ L \ \rightarrow \ |.$$

If we replace some of the symbols in this string by $\star$'s, the result will be a schema that is a building block for the condition. Examples of such replacements are 1$\star\star\star$ . . . $\star$, 10#$\star\star$ . . . $\star$, $\star$0$\star\star$0$\star$ . . . $\star$0$\star$, and $\star\star$ . . . $\star\star\star$10#. How many different ways can we insert $\star$'s in the given string? At each position we have two alternatives: we can either retain the symbol that is already there or we can insert a $\star$. So there are

$$2 \times 2 \times \ldots \times 2 = 2^L$$

$$| \ \leftarrow \ L \ \rightarrow \ |$$

different schemata for a single condition. For $L = 100$, there are

$$2^{100} \cong 10^{30}$$

schemata. This is an enormous number. If we were to calculate one million schema averages per second, it would still take longer than the estimated life of the universe to do one round of averages for all of the schemata *for a single condition*.

This leaves us with a considerable dilemma. It is not feasible to carry out the detailed calculations of schema averages that would let us conduct a detailed survey, and sophisticated analytic models provide limited guidance even in simple cases. What can we do?

GENETIC ALGORITHMS

For evolutionary processes, where there is no apparatus for calculating S-averages, the dilemma holds a fortiori. Yet the interaction of reproduction, crossover, and selection does discover and exploit building blocks. To give one example, the Krebs cycle is a useful building block, discovered early in evolutionary history, that has been used by a tremendous range of species. It is a basic eight-step metabolic cycle common to almost all cells that use oxygen, ranging from aerobic bacteria to humans. The genes that specify this cycle have almost identical alleles over this diverse range of cells. The Krebs cycle is just one example among many; any text on molecular biology will supply hundreds of other examples. It seems worthwhile to try to understand how evolution accomplishes this overwhelming computational task with no overt computational facility.

We can get a fairly accurate picture of what happens, even if we throw away most of the details. Simplify the whole reproduction cycle to consider only the reproduction and recombination of "chromosomes." Further simplify the process by representing the chromosomes as strings. Then use only two genetic operations: crossing over and mutation. Crossing over (*crossover*, for short) has already been described. Mutation, more precisely *point mutation*, is a process whereby individual alleles are randomly modified, yielding a different allele for the gene. In the rule strings, mutation could randomly flip a 1 at some position to a 0 or a #. In biological systems crossover is much more frequent than mutation, often as much as a million times more frequent.

To simulate the process of producing a new generation from the current one, we use the following three steps:

1. Reproduction according to fitness. Select strings from the current population (this might be the set of rules for the agent) to act as parents. The more fit the string (the stronger the rule), the more likely it is to be chosen as a parent. A given string of high fitness may be a parent several times over.

2. Recombination. The parent strings are paired, crossed, and mutated to produced offspring strings.

3. Replacement. The offspring strings replace randomly chosen strings in the current population. This cycle is repeated over and over to produce a succession of generations.

The key question is, what happens to building blocks (schemata) under this procedure? A bit of arithmetic is helpful here. To make things easy, let the fitness of a string directly determine the number of offspring it has in a given generation, and set the average fitness of the overall population to 1, so that the average individual produces 1 offspring. (None of this limits the validity of the point I want to make; it merely simplifies the calculations.)

Consider the building block 1★★ . . . ★ and for purposes of calculation assume it has just three instances in the population, with fitnesses 1, 0, and 1 respectively (see Figure 2.10). Let's see what happens to this building block under step (1). The three instances of 1★★ . . . ★ will produce a total of $1 + 0 + 1 = 2$ offspring, or an average of ⅔ offspring per instance. Note that this average is simply the average $S(1★★ . . . ★)$. Because these are the only strings carrying the building block 1★★ . . . ★, that building block will have only two instances in the new generation (assuming the parents persist for only one generation). Because $S(1★★ . . . ★) = ⅔$ is less than overall population average $A = 1$, this reduction in the number of instances of 1★★ . . . ★ is the outcome advocated by our earlier estimation procedure.

To see what changes when the numbers change, let's look at a second, more intricate building block in the same population. Consider ★0★#.#★★ . . . ★ and assume it also has three instances, with fitnesses 2, 2, and 1 respectively (see Figure 2.10). The three instances will produce a total of $2 + 2 + 1 = 5$ offspring, or an average of ⅗ offspring per instance. Again the outcome is just as the estimation procedure would advocate: $S(★0★#.#★★ . . . ★) = ⅗$ is greater than $A = 1$, so there should indeed be more instances of ★0★#.#★★ . . . ★ in the next generation.

We could repeat this calculation for each building block present in the population, obtaining in each case the outcome advocated by the estimation procedure: under reproduction according to fitness, above-average building blocks are used more frequently, and below-average are used less frequently.

For the mathematically inclined reader, this result can be given a succinct form. For any schema b belonging to $\{1,0,\#\}^L$, let $M(b,t)$ be the number of instances of schema b in the population at generation t. Then

$$M(b, \, t + 1) \, = \, S(b,t)M(b,t)$$

gives the number of instances in the next generation, at $t + 1$, after reproduction. Here $S(b,t)$ is the average strength of the instances of b at time t, already defined.

This is precisely the result desired, so why complicate the procedure by adding the crossover in step (2)? A moment's thought makes the reason obvious. The reproduction in step (1) simply copies strings already present; it does not produce any new combinations. In other words, step (1) does not produce any new hypotheses, so the agent would be limited to the best of the hypotheses present in the initial population. No matter how large the initial population, this can only be a minuscule sample of the possibilities. In a complex, changing environment, an agent using only step (1) is unlikely to fare well against agents that can generate new hypotheses. That is where crossover comes in.

EFFECTS OF CROSSOVER

Crossover can recombine schemata without greatly disturbing the desirable outcome of step (1). To see this, we have to take a more careful look at exactly what happens when two real chromosomes cross. The point at which they cross is not predetermined. In fact, the position at which the two cross over is about as likely to be one position as another (setting aside some skewing, caused by centromeres and other particular pieces of chromosomal apparatus). For present purposes we can assume that the point of crossover is chosen at random along the string.

What happens to a building block (schema) when the crossover in step (2) follows the reproduction of step (1)? We'll see that the effect depends on the *length* of the schema. That length is the number of possible crossover points between the outermost of the schema's defining positions (recall that a defining position is any position without a ⋆). For example, in the string ⋆0⋆#⋆⋆ . . . ⋆, positions 2, 4, and 5 are the defining positions, so the outermost defining positions are positions 2 and 5. There are three possible crossover points between these outermost positions, so the length of schema ⋆0⋆#⋆⋆ . . . ⋆ is 3.

Shorter schemata are less likely to be disrupted by crossover, because crossover cannot "break up" a schema unless it falls within the outer limits of the schema (see Figure 2.11). Schemata not broken up will be passed on to the next generation, as dictated by step (1). On a string of length L there are $L - 1$ possible points of crossover (the points *between* the genes). The chance of the crossover point falling within the outer limits of a schema is the length of the schema divided by $L - 1$. So in the example $\star 0 \star \# \# \star \star \ldots \star$, with $L = 100$, there are only 3 chances out of 99 that crossover will disrupt the schema. That is, 96 times out of 99 the schema will be passed intact to the next generation. The reasoning of step (1) holds.

In mathematical form, if $L(b)$ is the length of schema b, then $L(b)/(L - 1)$ is the probability that crossover will fall within the outer limits of b, and $1 - L(b) / (L - 1)$ is the chance that crossover will *not* fall within the outer limits of b. If we assume that every crossover falling within the outer limits actually disrupts the schema, then $1 - L(b) / (L - 1)$ is the chance the schema will *not* be disrupted. Accordingly, our earlier formulation, modified to take account of this effect of crossover, becomes

$$M(b,\ t + 1) = [1 - L(b) / (L - 1)]S(b,t)\ M(b,\ t),$$

where $M(b, t + 1)$ is the average or expected outcome because we are now dealing with a chance process, crossover.

Longer schemata, of course, have a much larger chance of being broken up; for a schema of length 50, when $L = 100$, crossover will fall within the outer limits more than half the time. There are two reasons why this disruption of longer building blocks is not much of a problem.

First, the above-average shorter schemata are the ones discovered early on. The reasoning is similar to that given for the early discovery of less specific conditions in default hierarchies: A schema must have one of the three letters {1,0,#} at each of its defining positions. Thus, if we select a particular set of k defining positions, 3^k variants are possible. For $k = 4$, there are therefore $3^4 = 81$ distinct schemata to be tested. Even a rather small population can, in a short time, have produced a useful number of trials of all of these alternatives. Because the number of defining positions for a schema is, at most, one more than its length, short schemata have fewer variants. These variants will be tested rather quickly, and if some are above average they will quickly be exploited, like the early exploitation of general rules in a default hierarchy.

Before we continue, it will be useful to recall the earlier point that

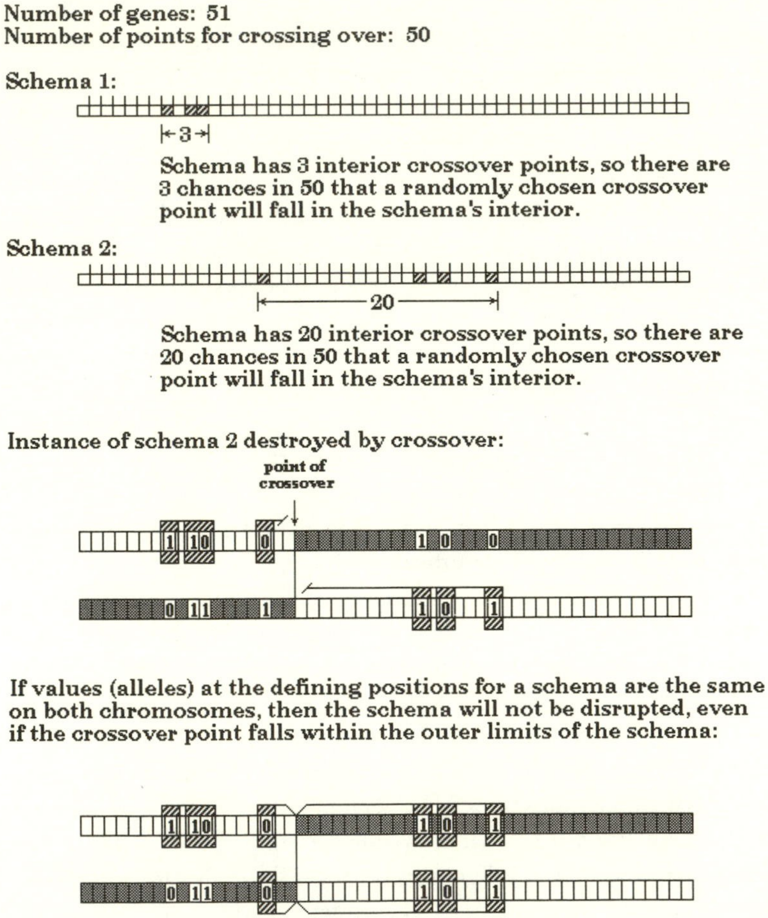

Number of genes: 51
Number of points for crossing over: 50

Schema 1:

|←3→|

Schema has 3 interior crossover points, so there are 3 chances in 50 that a randomly chosen crossover point will fall in the schema's interior.

Schema 2:

|←————— 20 —————→|

Schema has 20 interior crossover points, so there are 20 chances in 50 that a randomly chosen crossover point will fall in the schema's interior.

Instance of schema 2 destroyed by crossover:

point of
crossover

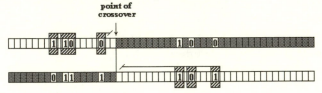

If values (alleles) at the defining positions for a schema are the same on both chromosomes, then the schema will not be disrupted, even if the crossover point falls within the outer limits of the schema:

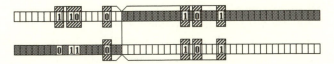

Figure 2.11   Effects of Crossover on Schemata.

there are approximately 10^{30} schemata present in a single string of length $L = 100$. Even if we limit ourselves to schemata defined on 4 positions, the number of such schemata present on a single string is still large. In fact, for $L = 100$ we can show that there are about 4,000,000 ways to choose different sets of 4 positions. (A simple calculation shows the number of distinct ways of choosing 4 things from a set of 100). Every single string contains each of these 4,000,000 distinct sets of 4 positions, so each string exhibits *one* of the 81 possible variants for *each* of those sets. Because there are only 81 alternatives for each set, we can still be assured that a rather small population will test *all* of the alternatives at *all* positions. Specifically, a population of a few hundred strings can produce useful estimates for all of the 81 × 4,000,000 schemata defined on 4 positions. A slightly more complicated calculation shows that even if these schemata are limited to a length of 10 or less, there are still more than 40,000,000 of them. Nine times out of ten, such schemata will be passed on to the next generation without disruption by crossover. Similar reasoning holds, of course, for other small numbers of defining positions.

From this we see that, with the genetic algorithm, the agent tests a very large number of schemata, even when we restrict our attention to the shorter schemata that are largely undisturbed by crossover. This is so even if the agent uses only a small number of rules (strings), because one rule in itself is an instance of a large number of short schemata, as we have just observed. It would be surprising if *none* of these short schemata were consistently associated with above-average performance.

The second reason that crossover's disruption of longer schemata is not so troubling stems from the observation that more complicated schemata are typically formed from combinations of shorter, well-established schemata. More complicated building blocks are usually formed by combining simpler building blocks. This fact reflects our earlier observation that innovations, such as the internal combustion engine, tend to involve a particular combination of relatively simple, widely used building blocks. Moreover, devices like the internal combustion engine become, in turn, the centerpiece of a wide range of still more complex devices. The result is a kind of hierarchy wherein the building blocks at one level are combined to form the building blocks at the next level. Under a genetic algorithm a similar hierarchy forms, wherein the higher-level (longer) schemata are typically composed of well-tested, above-average shorter schemata. This hierarchy ameliorates the disruptive effect of crossover, as we shall see very shortly.

First of all, under a genetic algorithm, above-average schemata soon come to occupy a large proportion of the population, because of above-average replication in step (1). Consider, then, two parent strings that contain identical copies of the same schema. Crossover cannot disrupt the schema, even if crossing over takes place inside the schema's outer limits. The alleles exchanged will be replaced by identical alleles (see Figure 2.11). It follows that crossover rarely disrupts longer schemata composed of particular combinations of shorter, above-average schemata. If some of these longer schemata are in turn above average, they spread through the population. The hierarchy becomes more elaborate, providing for the persistence of still longer schemata. A hierarchy of disruption-resistant schemata emerges, similar to the way default hierarchies emerge.

EFFECTS OF MUTATION

One question about step (2) remains. What is the role of mutation? To find out, we have to look to step (3), replacement. It is possible for a given schema, under reproductions, crossovers, and replacements ("deaths"), to come to be present in *every* member of the population. When this happens, all members of the population contain the same alleles at the positions on which the schema is defined. Say, for example, that the schema 1★★★ . . . ★ is present in all members, so that *all* strings in the population start with a 1. Then we have no strings that start with either a 0 or a #. In the set of all possible strings, $\{1,0,\#\}^L$, only ⅓ start with a 1. So, by losing just the two alleles 0 and # at position 1, we are reduced to trying out possibilities in only ⅓ of the space $\{1,0,\#\}^L$! Worse, once the alleles have been lost, the actions of reproduction and crossover cannot replace them. Under these circumstances the allele is said to have gone to *fixation*. If k alleles have gone to fixation, we are reduced to searching $(\frac{1}{3})^k$ of the space $\{1,0,\#\}^L$.

We might adopt the attitude that when an allele goes to fixation, the genetic algorithm has established that allele's superiority, so we need not try the alternatives any further. Unless we are very sure of the allele's superiority, this is a poor way to proceed. Our attitude has been one of sampling and estimation, because $\{1,0,\#\}^L$ is so large as to make it infeasible to try all combinations of alternatives. Estimates can be wrong, even after considerable testing. No matter how many trials underpin our estimate of the fitness of 1★★★ . . . ★, we cannot be sure that there is not a better string in the two-thirds of the space not being searched. This concern is particularly pressing when the value of a given building block (schema) depends on the context provided by other building blocks. It might be that the fitness of 0★★★ . . . ★ is vastly enhanced in the presence of ★11★#★★ . . . ★, and that we have yet to sample an instance of that combination. If the allele 1 at position 1 has gone to fixation, the genetic algorithm will have no chance to observe the combination of 0★★★ . . . ★ and ★11★#★★ . . . ★, unless the 1 at position 1 is driven away from fixation.

In mathematical form, if $P_{mut}(b)$ is the probability that a mutation will modify schema b, then $1-P_{mut}(b)$ is the probability that mutation will *not* modify b. Inserting this factor, as we did for crossover, we get

$$M(b,\, t+1) = [1-L(b)/(L-1)][1-P_{mut}(b)]S(b,t)M(b,t).$$

This formula, then, gives the number of instances of schema b we expect in the next generation after steps (1) and (2) of the genetic algorithm have been applied. This formula is, essentially, the *Schema Theorem* for genetic algorithms.

Mutation, by occasionally changing an allele to one of its alternatives, reopens the search. From time to time a 1 in the first position will be changed to a 0 or a #. In so doing, mutation provides the replacement that reproduction and crossover cannot. Calculations show that this "insurance policy" can be invoked with a mutation rate that is quite low compared to the crossover rate. This relation between mutation and crossover is in keeping with the fact that in biological systems mutation rates are orders of magnitude lower than crossover rates.

COMBINED EFFECTS

We can now put together all three steps of the genetic algorithm to see how they exploit above-average building blocks in producing a new generation. Step (1), reproduction according to fitness, causes *all* schemata to be treated according to the heuristic based on the estimation of schema averages: above-average schemata have more instances in the next generation, below-average schemata have fewer instances. In step (2), crossover generates offspring that are different from their parents, producing new combinations of the schemata passed on by step (1). Crossover sustains the increased use of shorter, above-average schemata but may disrupt longer schemata, particularly those not using shorter, above-average schemata as building blocks. Schemata not tried before may be generated by recombination of fragments when crossover disrupts extant schemata. That is, crossover may generate new schemata even as it recombines those already present. Mutation acts in step (2) to provide an insurance policy against loss of alleles, and it can also generate new schemata by altering the defining positions of extant schemata. Finally, in step (3), the offspring replace strings already in the population. This process introduces a "death rate" just sufficient to keep the population at a constant size. These combined effects are summarized in mathematical form by the Schema Theorem (in a form closely related to the equation at the end of the previous section).

The most important feature of a genetic algorithm is its ability to carry on this sophisticated manipulation of building blocks by acting only on whole strings. We saw earlier that the number of building blocks is so large that it is not feasible to calculate explicitly the estimates of schema fitness that would guide increased or decreased usage of given building blocks. The genetic algorithm does implicitly what is infeasible explicitly. The whole-string operations (reproduction, crossover and mutation) do not directly deal with schemata and carry out no computations involving them. Yet the algorithm acts as if such computations were being made and exploited. Above-average schemata of one generation are used more frequently in the next generation and below-average schemata are used less frequently. This ability to manipulate large numbers of schemata implicitly through the explicit manipulation of a relatively small number of strings is called *implicit parallelism*.

Viewing rule discovery in terms of building block manipulation and implicit parallelism changes the outlook in another way. Consider a biological population, say a human population. No individual in a given generation is identical to any individual of the previous generation. Even the best individuals in a generation are never repeated in a future generation. There will only, ever, be one Einstein. Here we have a bit of a dilemma. If evolution "forgets" the best individuals in each generation, what does it "remember"? Implicit parallelism supplies an answer. Particular individuals do not recur, but their building blocks do.

This recurrence of building blocks is a familiar feature of artificial breeding. Every thoroughbred breeder knows that certain desirable features are associated with particular bloodlines. These are the building blocks that are combined by selective crossbreeding. Though we will never again see Man o' War or Citation, their building blocks will appear again and again.

Evolution "remembers" combinations of building blocks that increase fitness. The building blocks that recur generation after generation are those that have survived in the contexts in which they have been tested. These contexts are provided by (1) other building blocks and (2) the environmental niche(s) the species inhabits. There is actually an extensive hierarchy that is continually tested at every level. At the lowest level are particular, short DNA sequences that provide standard tags. These help implement the DNA translation process, such as the "start" and "stop" codes for translation of the DNA sequences that make up the chromosome's alleles. At the next level are the alleles themselves, and one level above that are combinations of alleles, the *coadapted* alleles, that code for enzymes that work well together. The Krebs cycle is an example of such a coadapted set, remembered over hundreds of millions of years.

The building blocks that we observe are, by and large, the *robust* building blocks. The Krebs cycle is so robust that it occurs throughout whole kingdoms of organisms. Under this view, evolution continually generates and selects building blocks at all levels, selected combinations of established building blocks at one level becoming the building blocks of the next-higher level. Evolution continually innovates, but at each level it conserves the elements that are recombined to yield the innovations. When a new building block is discovered at some level, it usually opens a whole range of possibilities because of the potential for new combinations with other extant building blocks. Tremendous changes and advances ensue. The genetic algorithm, applied to rule discovery, mimics this process but with a much simpler syntax.

An Example: An Adaptive Agent for the Prisoner's Dilemma

The Prisoner's Dilemma is a two-person game that captures major political and personal interactions in a simple, well-defined context. The interested reader can learn about the history and importance of this game in Axelrod (1984). The game is of particular interest because the solution given by the theory of games is to avoid cooperation (called defection) whereas, in actual repeated play, players discover the benefits of mutual cooperation. Let me describe the game in greater detail, then show how adaptive agents learn to play.

In the Prisoner's Dilemma, each player has just two options at each play, known colloquially as "cooperate" (C) and "defect" (D). There are, therefore, four possible outcomes to a given play of the game: (C,C), both players elect to cooperate; (C,D), first player cooperates and second player defects; (D,C), first player defects and second player cooperates; and (D,D), both players defect. The payoff (relative value) of these outcomes is given by the following table:

| | | Second Player ||
		C (cooperate)	D (defect)
First Player	C (cooperate)	+3, +3	0, +5
	D (defect)	+5, 0	+1, +1

For instance, the outcome (D,C) is worth +5 to the first player and 0 to the second, as given by the pair (+5, 0) in the table.

The minimax solution given by game theory minimizes the maximum damage the opponent can do. It is determined by comparing the maximum damage under cooperation with the maximum damage under defection. If the first player cooperates (C,-), the maximum damage occurs when the second player makes the response D, yielding (C,D) with a payoff of 0 to the first player. If the first player defects (D,-), then the maximum damage again occurs when the second player makes the response D, but now the payoff to the first player is +1. Thus, the first player suffers minimum damage by always defecting. The same reasoning holds for the second player. Thus (D,D) is the minimax solution.

It is evident from the table that both players can do better. If they can come to mutually cooperate (C,C), both can earn +3 on each play, a much better outcome than the minimax solution. In actual repeated play, players discover the benefits of mutual cooperation after trying out various strategies, and the game typically settles down to long bouts of cooperation. Experiment shows that a quite simple strategy, *tit for tat*, induces cooperation while punishing defection. To understand this strategy, we need to know more about the notion of a strategy for playing the Prisoner's Dilemma.

A strategy for repeated play of the game uses the recent history of play to choose one of the two options for the *next* move. Here, we simplify by setting a "horizon" so that each player can only remember the past three outcomes. At time t, then, the history would be the outcomes at $t-3$, $t-2$, and $t-1$, say (C,D), (C,D), and (D,D). With this horizon there are $4 \times 4 \times 4 = 64$ possible distinct histories ranging from (C,C) (C,C) (C,C) to (D,D) (D,D) (D,D). They are listed in the history column of the table below. A strategy must specify, for *each* history, what move (C or D) the player should make.

The table presents a particular strategy, tit for tat. The reply of the first player at time t ("tit") simply duplicates the action of the second player the previous time, $t-1$ ("tat"). When a history ends in D, therefore, the next action taken should be D, whereas if it ends in C, the action taken should be C.

We can assign each of the 64 histories an index. Assign index 1 to history (C,C) (C,C) (C,C) and index 64 to history (D,D) (D,D) (D,D). Thus (from the table), the strategy might say under history 1, (C,C) (C,C) (C,C), cooperate (C); under history 2, defect (D); and so on through history 64.

Index	History			Action
	t-3 *t*-2 *t*-1			*t*
1	(C,C) (C,C) (C,C)			C
2	(C,C) (C,C) (C,D)			D
3	(C,C) (C,C) (D,C)			C
4	(C,C) (C,C) (D,D)			D
•	•			•
•	•			•
•	•			•
64	(D,D) (D,D) (D,D)			D

Using the indexes for histories, a complete strategy can be represented by a string with 64 positions. At the first position in the string we insert the action to be taken under history 1, at the second position the action to be taken under history 2, and so on.

Index (histories):   1 2 3 4   • • •   64
String (actions):   C D C D   • • •   D

The tit-for-tat strategy, then, places a C at the odd-numbered positions and a D at the even-numbered positions, yielding the string

CDCDCDCDCDCDCDCDCDCDCDCDCDCDCDCDCDCDCDCDCDCDCDCDCDCDCDCDCDCDCDCDCD.

A quick calculation shows that even for a game as simple as the Prisoner's Dilemma with a three-step horizon, the number of possible strategies is overwhelming—2^{64}, which is approximately equal to 16 billion billion!

We can think of a player learning to play the repeated Prisoner's Dilemma by starting off with a small *set* of sample strategies to be tested against the opponent. We can also think of each strategy as a set of stimulus-response rules, where the immediate past history is the stimulus that determines which play is to be made in response. Adaptation, then, involves (1) assignment of ratings to each of the strategies on the basis of experience, and (2) invention of new strategies to replace those that end up with low ratings. The rating of a strategy is merely the average of the payoffs it receives when it is used against the opponent. The genetic algorithm uses these ratings as fitnesses and generates new strategies accordingly.

Repertoire of Strategies (player A)	Average Payoff (against player B)
CDDCCCCDCCDCDDDCDDCCDCD . . . DCDCCDCCC	+0.5
DDDDDDDDDDDDDDDDDDDDDDDD . . . DDDDDDDDD	−0.4
CCCCDDDDCCCCDDDDCCCCDDD : . . DCCCCDDDD	+0.7
.	
.	
.	
CCCCCCCCCCCCCCCCCCCCCCCC . . . CCCCCCCCC	−0.2

It is an interesting sidelight that we can anticipate what schemata (building blocks) will be used by the genetic algorithm, because we know that tit for tat is a favorable strategy. C's at even positions and D's at odd positions are components of a tit-for-tat strategy, so that combinations of C's and D's satisfying this requirement should enhance performance. For example, the combination CDCD placed so that the C falls at an odd position would be a useful schema. According to the schema theorem for genetic algorithms, such building blocks should appear ever more frequently as new strategies (strings) are generated.

Moreover, as building blocks at different positions become common, crossover can combine them, providing offspring with still more building blocks.

Parent Strings	Offspring Strings
crossover point	
DDD CDCD DCCCDDDDDCC ... CCCCC	DDD CDCD DCCC CDCDCD D ... CDDDC
CCDDDCDDD CC CDCDCD D ... CDDDC	CCDDDCDDDCCDDDDDCC ... CCCCC

Robert Axelrod at the University of Michigan, with the help of Stephanie Forrest, designed a simulated player that started with a small set of randomly chosen strategies (see Axelrod, 1987). The simulated player employed the genetic algorithm to search the large set of possible strategies. The hope was that the genetic algorithm would find the tit-for-tat strategy after a reasonable number of plays. In fact, the genetic algorithm did more than that. After discovering tit for tat, it actually generated a strategy better than tit for tat. This strategy exploited players that could be "bluffed," reverting to tit for tat when history indicated the player could not be bluffed.

Adaptive Agents and Economics

That adaptive agents can learn strategies in a game like the Prisoner's Dilemma, combined with the close relationship between games and economics, suggests an approach to economics based on adaptive agents. Conversations with Brian Arthur at the Santa Fe Institute induced me to pursue thoughts along these lines in a more than casual way. Our ideas, encouraged by interplay at some seminal workshops at the institute set in motion by Philip Anderson and Kenneth Arrow, solidified into a project for simulating a stock market using adaptive agents. This project was to be a thought experiment, not an attempt at prediction; it was aimed at getting a better feeling for the dynamics of the market.

Though it might seem otherwise, market dynamics are not a natural area of study for classical economics. From the classical point of view, markets should always clear rapidly, moving in narrow ranges dictated by changing supply and demand. Classical models do not readily generate crashes and speculative bubbles. It is easy to pinpoint the reason for this lack. Classical theory is built around agents of perfect rationality—agents that perfectly foresee the consequences of their actions, including the reactions of other agents. Unusual dynamics, such as crashes and speculative bubbles, are usually attributed to incidentals, such as noisy degradation of information.

Still, real markets typically fluctuate much more rapidly, and over much wider ranges, than the supply and demand fluctuations that supposedly drive them. Both Arthur and I felt that a market based on adaptive agents, agents of bounded rationality rather than agents of perfect rationality, was much more likely to exhibit "natural" dynamics. In particular, we felt that the anticipatory speculations produced by the internal models of such agents would generate speculative bubbles and subsequent crashes. In other words, we felt that learning, and the imperfect internal models it produces, would automatically generate realistic dynamics without the introduction of exogenous variables. With a computer-based model, we could see just how far the mechanisms of the adaptive agent syntax would take us.

We proceeded to implement this approach, recruiting others, such as the physicist Richard Palmer, as we went along. In our model a small number of adaptive agents trade in a single stock, with a (non-adaptive) specialist program adjudicating buy and sell offers to determine a current price (the equivalent of a daily average). To produce the "anonymity" of the stock market, and to keep things simple, an agent's only input information on each time-step is this current price. On the basis of this information, perhaps collected in a "history" (as in the Prisoner's Dilemma), the agent decides on one of three actions at each time-step: BUY, SELL, or HOLD. There is a "dividend" on shares held, so that an agent makes money by simply holding. (This dividend, which does not fluctuate in the simplest models, determines a "fundamental value" for the stock.) The measure of performance of any given agent is the amount of money it accumulates through its actions.

The details of this implementation do not add much to the description just given, and the earlier example of the Prisoner's Dilemma gives some idea of what is involved. So I'll go directly to results. In a typical run, the agents are started with randomized initial strategies. As might be expected, the initial market is pretty disorderly. But credit assignment and the genetic algorithm soon provide each individual agent with experience-based rules for buying, selling, and holding. An agent might develop rules of this form: IF (the price is 40) THEN (sell), and IF (the price is below 40) THEN (buy). The market soon smooths out and begins to look like a market involving the agents of classical economics. Then one of the agents finds a rule that exploits the market's "inertia," making money by selling a bit "late" in a rising market. Other agents begin to anticipate trends, and the whole learning process yields a market which makes these trend projections self-fulfilling—for a while. Over time, after enough self-fulling prophecies, the behavior becomes more and more exaggerated, leading to a bubble and eventually a crash. The whole process seems quite natural, and not the least surprising, in this framework. When we "dissect" the agents, we even find sets of rules that mimic, in this simple setup, well-known market strategies such as "chartism."

Ours is not the only computer-based model using adaptive agents to emerge from the Santa Fe Institute workshops in economics. Another model, every bit as interesting as the stock market model, was designed by Ramon Marimon and Thomas Sargent (see Marimon, McGratten, and Sargent, 1990). This model is built on Wicksell's Triangle, a classic model in economics. Wicksell's Triangle consists of three "countries," each of which produces a single product. A problem arises because the product a country produces is not the product it wants to consume; the product it desires is produced by one of the other countries. What is an efficient trading pattern for these countries? Among other things,

Wicksell's Triangle concerns the emergence of "money," the use of one of the products as a medium of exchange.

The scope for action of each of the countries in Wicksell's Triangle is so simple that it seems readymade for a computer-based simulation based on adaptive agents. The triangle has been much studied by economists, so that various mathematical approaches are available for comparison. The simulation, starting with randomly endowed agents, did exhibit the emergence of one product as a medium of exchange under a wide variety of conditions. In the simulation the conditions for emergence were examined in some detail, providing guidelines for determining which of the products would serve as a basis for other exchanges.

These early efforts at using adaptive agents to study bounded rationality, and the ensuing dynamics of economies so described, seem to me suggestive and hopeful. Because such systems do not settle down, or even stay at a quasi-equilibrium for long, they provide a window on aspects of economics not often available for rigorous study. An economist may ask, "What can we study in a system that exhibits perpetual novelty?" But the situation is not so different from that faced by a meteorologist. On all time and distance scales, the weather goes through never-repeating changes. While we cannot predict weather in detail over more than a few days, we understand the relevant phenomena well enough to make many useful adjustments, both short term and long term. For our adaptive-agent-based studies of economies, we must find the counterpart of fronts and jet streams (tagged aggregates, mind you) if we are to make progress. Then we may be able to uncover some of the critical lever points.

Recapitulation

We can now step back to see just what we've given up and what we've retained in this framework for representing adaptive agents. The framework, as intended, consists of three major components: (1) a performance system, (2) a credit-assignment algorithm, and (3) a rule-discovery algorithm.

(1) The performance system specifies the agent's capabilities at a fixed point in time—what it could do in the absence of any further adaptation. The three basic elements of the performance system are a set of detectors, a set of IF/THEN rules, and a set of effectors. The detectors represent the agent's capabilities for extracting information from its environment, the IF/THEN rules represent its capabilities for processing that information, and the effectors represent its ability to act on its environment. For all three elements the abstraction loses the details of the mechanisms employed by the different kinds of agents.

A closer look at the concept of detectors gives us a better idea of what has been lost and what has been gained. An antibody employs detectors that depend on local arrays of chemical bonds, while an organism's detectors are best described in terms of its senses, and a business firm's detectors are usefully described in terms of the responsibilities of its various departments. In each instance there are interesting questions about the particular mechanisms for extracting information from the environment, but we have put these questions aside here. Our framework concentrates on the information produced—the properties of the environment to which the agent is sensitive. We exploit the fact that any such information can be represented as a binary string, here called a message. We gain the ability to describe, in a uniform way, any agent's ability to extract information from its environment. Defining the performance system's ability to affect the environment in terms of message-sensitive effectors entails similar losses and gains.

The same considerations hold for the agent's ability to process information internally. The mechanisms are various, but we have concentrated on the information-processing aspect. By conjoining IF/THEN rules with messages, we wind up with rules of the form IF (there is a message of type c on the message list) THEN (post message m on the list). In so doing, we lose the details of the mechanisms used by particular agents for processing information. For example, if we are studying the progression in which genes are turned on and off in a developing embryo, we lose all the fascinating details about the particular mechanisms of repression and derepression. We retain, however, a description of the stages of development, and the information fed back at each stage. In general, we gain the ability to describe any information-processing capability that can be modeled on a computer. Because many rules can be active simultaneously, we gain a natural way for describing the distributed activity of complex adaptive systems. In particular, systems with this parallelism automatically describe novel situations in terms of familiar components; internal models, in the form of default hierarchies, form naturally. Both activities are pervasive in *cas*.

Once we settle on a rule-based description of performance, the process of adaptation provides components (2) and (3) of the framework.

(2) The essence of credit assignment is to provide the system with hypotheses that anticipate future consequences—strengthening rules that set the stage for later, overtly rewarding activities. For *cas* this process leads to a question we have not really explored so far. Just what is it that should be considered rewarding? We'll look at this question in some depth in the next chapter, but let me touch on it here.

In mathematical studies of genetics, economics, and psychology this question is often settled by fiat, assigning numerical values to the objects of interest. Fitness is directly assigned to chromosomes, utility is directly assigned to goods, and reward is directly assigned to behaviors. But the question is more subtle. Consider the behavior of an organism. Generally, evolution has built in certain internal detectors that record the status of "reservoirs" of food, water, sex, and the like. The organism's behavior is largely directed at keeping these detectors away from "empty." For more sophisticated organisms, much stage setting and anticipation goes into this task. It is a kind of never-ending game with intermittent payoffs. The value of any behavior depends on the current position in the game and the status of the reservoirs. Said another way, figures of merit for *cas* are usually implicitly defined. Competition, with local payments, is one of the few techniques we have for handling such problems in distributed systems. We'll soon see how pervasive such competition is in *cas*; for now we simply note that competition is the basis of the credit-assignment technique used to describe this aspect of adaptive agents.

(3) Rule discovery, the generation of plausible hypotheses, centers

on the use of tested building blocks. Past experience is directly incorporated, yet innovation has broad latitude. This particular method of recombining building blocks draws heavily on genetics, but it can be considered as an abstract version of a pervasive process. We can even describe neurophysiological theories of thought in terms of building blocks. Take Hebb's (1949) classic, still influential treatise. In Hebb's theory a *cell assembly* is a set of a few thousand interconnected neurons capable of self-sustained reverberation. A cell assembly operates somewhat like a small cluster of rules that is coupled via common tags. Cell assemblies act in parallel, broadcasting their messages (pulses) widely via a large number of synapses (interneuron contacts—a single neuron may have as many as ten thousand synapses). Cell assemblies compete for neurons via recruitment (adding parts of other cell assemblies) and fractionation (dividing into fragments that serve as offspring). It is not difficult to see this as a process that recombines tested building blocks. Moreover, cell assemblies can be integrated into larger structures called *phase sequences*. Indeed, it is not difficult, on rereading Hebb, to see counterparts of all the processes we have discussed.

Because tags play such an important role in coupling rules and providing sequential activity, it is important to note that they too have building blocks. Tags are really schemata that appear in both the condition and action parts of rules. As such, they are subject to the same manipulations as any other part of a rule. Established tags—those found in strong rules—spawn related tags, providing new couplings, new clusters, and new interactions. Tags tend to enrich internal models by adding flesh (associations) to the skeleton provided by default hierarchies.

Onward

With these definitions and procedures in place we have a uniform way of depicting the daunting array of adaptive agents that appear in *cas*. The availability of a uniform description for adaptive agents gives hope that we can indeed portray all *cas* within a common framework. Cross-comparisons of different *cas* then take on added meaning because they can be made in a common language. We can translate mechanisms that are salient and obvious in one *cas* to other *cas* where the mechanisms may be obscure, though important. Metaphors and other guides in the search for general principles become enriched. The search becomes more directed, and more hopeful.

To see where this may lead, look again at New York City. Interesting comparisons are possible even when the systems are at opposite ends of the *cas* continuum. Consider an embryo as the metaphorical counterpart of the city. If we look to the origins of New York four centuries ago and make appropriate changes in timescale, the growth of the city does show some similarity to the growth of an embryo. Both start from a relatively simple seed. Both grow and change. Both develop internal boundaries and substructures, with a progressively more complicated infrastructure for communication and transport of resources. Both adapt to internal and external changes, retaining coherence while holding critical functions in narrow ranges. And, underpinning all, both consist of large numbers of adaptive agents—in one case, various kinds of firms and individuals, and in the other, a variety of biological cells.

Can we make these similarities into something more than an interesting anecdote? Are there lever points of embryonic development (and we know quite a few from work in morphogenesis, for example; see Buss, 1987) that are suggestive in altering urban development? Later we'll see that crises offer unusual opportunities for changing urban habits. Are the experimental crises we induce in embryos suggestive in this respect? Can we make comparisons in "anatomy" that will be helpful in the way that Darwin's anatomical comparisons enabled him to advance the theory of natural selection?

To make progress on this and similar questions, we need to use our common representation for adaptive agents in a broader setting. We have to provide an environment that allows our genetic agents to interact and aggregate. That is the subject of the next chapter.

· 3 ·

Echoing Emergence

W<small>E CAN NOW DESCRIBE</small> the actions and interactions of an adaptive agent in some detail, and we can do so in a common format, whatever the agent's outward form. With our new understanding of the process of adaptation as background, it's time to look at complex adaptive systems as a whole. Here we confront directly the issues, and the questions, that distinguish *cas* from other kinds of systems. One of the most obvious of these distinctions is the diversity of the agents that form *cas*. Is this diversity the product of similar mechanisms in different *cas*? Another distinction is more subtle, though equally pervasive and important. The interactions of agents in *cas* are governed by anticipations engendered by learning and long-term adaptation. In specific *cas*, some anticipations are held in common by most agents, while others vary from agent to agent. Are there useful aggregate descriptions of these anticipations? The combination of diversity and anticipation accounts for much of the complexity of *cas* behavior. Both seem to arise from similar mechanisms for adaptation and evolution. Is there a way to weld these mechanisms into a rigorous framework that encompasses all *cas*?

It is only through a unifying model that we can develop a deeper understanding of such critical phenomena as the lever-point phenomenon. We know specific examples of this phenomenon: the vaccines that act as levers on the immune system, the enzymes that direct and redirect activities in the cell, the sudden fright that permanently changes the central nervous system, the introduction of an organism (say a rabbit) into an ecosystem where it has no natural enemies (Australia), and so on. There even seem to be similarities among these examples. But we're far from characterizing the conditions in *cas* that make leverage possible. If we look to a different *cas*, the search begins anew, with no help from previous instances. The examples just given tell us little about the kinds of economic conditions that encourage the tremendous growth and financial leverage of a Microsoft Corporation. We need guidelines that go beyond specific *cas*, and we're likely to find them only when we understand the general principles that underpin *cas*. That understanding, in turn, is likely to arise only with the help of computer-based models that extract the essence of *cas*.

The attempt to provide a framework and theory that applies to all *cas* depends, as is usual in the sciences, on two activities: (1) the provision of an organized set of data, and (2) the use of induction, aided by mathematics, to find laws that can generate those data. This is a familiar process, often described in textbooks, but it helps to have a canonical example. One of my favorites comes from the early days of science. Tycho Brahe, as part of his extensive efforts in the sixteenth century, kept a careful record of the nightly positions of the planets, which over the course of months move through the skies in a kind of S-shaped curve. Later, after an extended search, Kepler produced the insight that ellipses, with the sun at one focus, can generate those data. (The interaction between Brahe and Kepler, and the scientific results, are nicely described in Lodge, 1950.) When this classic process is translated to the study of *cas*, we'll see that it takes some unusual twists.

The present chapter uses a series of increasingly complex models to illustrate the process of selection and rejection that goes into organizing complex data. I worked on an early precursor of these models in 1975 (Holland, 1976), and some of the ideas were honed in a seminar organized by Doyne Farmer and Chris Langton during my year on "the Hill" (Los Alamos National Laboratory) as Ulam Scholar. However, the spark that directly ignited the work was a request from Murray

Gell-Mann: he asked if I could produce a simple, highly visual model that would illustrate the creation of complex structures by natural selection. It is difficult to say no to Murray, and he is persistent. I began to think of ways to satisfy his request while furthering my own research objectives. The Echo model is the result, though I fear it does not yet meet Murray's needs.

Echo relies on the basic mechanisms and properties enumerated in Chapter 1 to provide a framework for examining *cas*. By turning this framework into a computer-based model (the subject of the next chapter), we attain a fully rigorous presentation. The computer-based version can be "run," so that we can observe the actions of its mechanisms and the resulting behavior. (It is rather as if Brahe and Kepler had a mechanized orrery for generating the positions of the planets.) Because *cas* are so intricate, computer-based models, with their well-defined, manipulatable mechanisms, provide a crucial intermediate step in the search for *cas* laws. Such models, where they mimic relevant *cas* phenomena, put *cas* data in a rigorous format, thereby facilitating the description of patterns and laws.

Organizing Cas Data

Organizing data can sometimes be simple. Brahe merely recorded time and position for each planet. It becomes difficult when there are many things that *could* be recorded. The modern experimental physicist thinks long and hard about what instruments or gauges to use and under what conditions. And these thoughts are guided by what theory suggests should happen, or by holes in current theory. If the experimenter is inspired, the result is a *critical experiment*, where some assumed law or mechanism is shown to be adequate, or inadequate, to generate selected data. In setting up the experiment, the researcher decides what is to be included and what is to be excluded, as well as what is to be held constant (if he or she has that much control). The experimenter does much to organize the data merely by organizing the conditions of the experiment.

Cas present substantial problems when it comes to extracting and organizing data. As with astronomy, the experimenter cannot stop the system in order to run it again under different conditions. He or she may even be constrained in the ways the system can be probed. An economist may be reasonably sure that high interest rates discourage long-horizon investments, but it is not an experiment that will be tried under controlled conditions, even if the economist has the power to carry it off. All too often *cas* seem to adhere to a version of the "Third Harvard Law of Biology": with a careful research plan, under controlled conditions, using selected agents, complex adaptive systems do pretty much as they damn please.

At the start of this book, I emphasized that, in building models, we must distill pervasive characteristics from idiosyncratic features. This point holds a fortiori when we are trying to develop models that are relevant for all *cas*. It is a more than usually difficult task for *cas*, because these idiosyncratic features are often a fascinating and diverting subject in themselves. However, our hope for a general understanding depends on setting them aside. We need the distillate—simpler models that supply guidelines for the study of all *cas*.

Computer-based models help because they can be started, stopped, and manipulated to one's heart's content. This very flexibility is a source of difficulty, though. A computer-based model is already an abstraction from data, even when it is designed to carefully mimic a specific system. Of course, this is also more or less true of a carefully designed physical experiment—such an experiment does deal directly with physical objects, but many influential factors have been deliberately damped down or excluded. The computer-based model goes farther down this path. At no point is it automatically constrained by physical reality. The experimenter can impose any computable laws, and they can be as fanciful as desired or accidentally permitted. Caution and insight are the watchwords if the computer-based model is to be helpful.

Even a model designed for thought experiments must still attend to data or laws derived from data. The designer must still carefully select the setting, as with a physical experiment. But there is the added constraint that the setting must be physically plausible, a condition automatically met in the physical experiment. A model does organize data, and in this it is like the table Brahe used for that purpose; but a computer-based model does more. When the model is run, it rigorously unfolds the consequences of its design (Brahe's tables become active!). This activity turns the computer-based model into a halfway house between experiment and theory. Looking back to data, we can see if the consequences are plausible; looking forward to theory, we can see if general principles are suggested.

Discovering lever points and other critical *cas* phenomena is particularly difficult because contexts and activities are continually changing as the agents adapt. It is rare that we can even determine the utility of a given activity. The utility of the various activities of a given agent depends too much on the changing context provided by other agents. In mimicry, symbiosis, and other properties, the welfare of one agent depends critically on the presence of other, different agents. Fitness (reward, payoff) is implicitly defined in such cases. We cannot assign a fixed fitness to a chromosome because that fitness, however defined, is context dependent and changing. So it is for all *cas*. Our first order of business, then, is to provide a class of models in which the welfare of an adaptive agent stems from its interactions rather than from some predetermined fitness function.

We are entering new territory. Few models exist that exhibit this implicit approach to fitness, even in quite simple situations. There is more of a mystery to the origin of the pin factory that Adam Smith (1776) discusses in his *Wealth of Nations* than is generally realized. This factory was one of the first examples of a production line; one craftsman drew the wire, another clipped it to size, another sharpened the point, and so on. The result was a tenfold increase in production over the efforts of the same number of craftsmen working individually. Smith and later commentators discuss relevant factors: specialization, more efficient learning, mass purchasing, and so on. But we do not have any models that demonstrate the transition that enables individual skilled craftsmen to organize into a factory. What actions and interactions between these individual agents produced an organized aggregate that persisted? What were the adaptive mechanisms that favored the emergence of this aggregate? It makes little sense, and it helps our understanding not at all, to assign a high a priori fitness to the pin factory. That fitness must emerge from the context.

The Criteria for Echo

At this point, we need a concrete example of the kind of model I've been describing. To that end, I'll devote the rest of this chapter to the formulation of such a model, really a class of models, called *Echo*. By illustrating both the possibility and the possibilities of a unifying model, Echo gives us a way of rephrasing the questions we've encountered so that they apply to all *cas*. Echo has been formulated with several criteria in mind:

(1) Echo should be as simple as possible, consistent with the other criteria. It is meant for thought experiments rather than for emulation of real systems. (Despite the simplicity, it can actually be used to model some real experiments, a case in point being Brown, 1994—data about the ongoing changes in an ecosystem in Arizona when a major predator, the kangaroo rat, is excluded from the system.) This simplicity is attained, in part, by substantially restricting the latitude of the adaptive agents in Echo. Interactions are carefully constrained, and the agents have only primitive internal models.

(2) Echo should be designed so that the actions of its agents are interpretable in a wide range of *cas* settings. In particular, the model should provide for the study of interactions of agents that are distributed in space (a "geography") and are mobile. It should be possible to assign different inputs (stimuli and resources) to different sites in the space when desired.

(3) Echo should facilitate experiments on the evolution of fitness. To this end, fitness in Echo should *not* be fixed at the outset as something outside the system (an exogenous factor). Rather, fitness should depend on the context provided by the site and other agents at that site (endogenous factors). The fitness of an agent should change as the system evolves.

(4) The primitive mechanisms in Echo should have ready counter-

parts in all *cas*. Two advantages follow. Interpretations of the results are constrained to be consistent with the ready-made interpretations of the mechanisms. Simulations, after all, are simply manipulations of numbers and symbols. It is all too easy to label output in facile, even fanciful, ways, thereby giving an "eye-of-the-beholder" distortion to the interpretation. The grounding provided by the interpretations of the primitive mechanisms counters this tendency by constraining the labeling. A second advantage accrues because, with the help of the interpretations, selected mechanisms can be shown to be sufficient to generate the phenomena of interest. In evolutionary biology, for example, there has been an extended discussion about the sufficiency of standard Darwinian mechanisms as a means of generating the saltations that appear in the paleontological record (see Gould, 1994). While simulations cannot establish that a given mechanism is actually present—only observation can do that—they can establish the sufficiency or plausibility of the mechanism.

(5) The Echo models should be designed to incorporate well-known models of particular *cas* wherever possible. This is a version of the Correspondence Principle that Niels Bohr applied so effectively to the development of quantum physics (see Pais, 1991). There are well-studied mathematical models that apply to all *cas* when suitably translated: biological arms races (Figure 1.12 and Dawkins, 1976) and survival of mimics (Brower, 1988) in ecology; Wicksell's Triangle (Marimon, McGratten, and Sargent, 1990), and Overlapping Generation models (Boldrin, 1988) in economics; the Prisoner's Dilemma game (Axelrod, 1984) in political science; Two-Armed Bandits (Holland, 1992) in operations research; and antigen-antibody matching in immunology (Perelson, 1994). If we can incorporate these translations in the Echo framework as special cases, we gain several advantages. We make bridges to paradigmatic models that have undergone intense scrutiny in the disciplines in which they originated—they have already been adjudged to be useful abstractions of critical problems. When Echo incorporates these abstractions as special cases, it benefits from the thought and selection that went into them. As another benefit, Echo becomes more accessible, and more open to critical inspection, in the

originating disciplines. Also, as with the interpreted primitive mechanisms, these abstractions ground Echo more firmly, constraining eye-of-the-beholder interpretations.

(6) As many aspects of Echo as possible should be amenable to mathematical analysis, the surest route for arriving at valid generalizations from specific simulations. The Bohr-like correspondences should supply mathematical landmarks that we can link into a more complete map, under the guidance of simulations.

In developing a version of Echo that meets these six criteria, I've taken a step-by-step approach rather than try to go directly to a single overarching model. Each step adds one additional mechanism, or modification, then describes what is gained thereby. Even the first model in the progression meets all the criteria to some degree. It places particular emphasis on avoiding an overt fitness criterion: agents live or die in terms of their ability to collect critical resources. As further mechanisms are added, the means for collecting critical resources expand. Counterparts of predation, trade, scavenging, specialization, and so on all can arise and evolve significantly as the agents evolve. Any combination of the primitive mechanisms that provides adequate amounts of resources for the agent, however bizarre, is passed on and becomes a building block for future generations. The last model in the sequence looks to the changing fitness of agents having increasingly diverse organizations, including structures that develop from seedlike founders.

Only the first model in this sequence has undergone extensive testing, though relevant parts of the others have been simulated. It will be easier to discuss what has been left out, and what remains to be done, after I have described the models. The last section of this chapter provides a scenario of the interactions that the most sophisticated model is designed to exhibit. As various levels are tested, we should gain useful guidelines for investigating real *cas*, even if only a few of the anticipated interactions show up. In this the models have a role similar to mathematical theory, shearing away detail and illuminating crucial features in a rigorous context. They differ from mathematics in that they do not rigorously establish generalizations.

The Organization of Echo

RESOURCES AND SITES

Echo's foundation is laid by specifying a set of "renewable" *resources*, which are treated quite abstractly. They can be represented by letters so that, for example, we might have four resources symbolized by the letters $\{a,b,c,d\}$. *Everything* in Echo is constructed by combining these resources into strings. The resources are treated much like atoms, being combined into "molecular" strings. However, no sophisticated bonding properties are associated with the resources; all strings are admissible. Thus, with $\{a,b,c,d\}$ as resources, any string based on these four resources, such as *ab*, or *aaa*, or *abcdabcd*, would be an admissible structure in Echo. We'll see shortly how agents are constructed from these strings.

Echo's "geography" is specified by a set of interconnected *sites* (see Figure 3.1). The neighborhood relation between sites—the pattern of juxtapositions—can be quite arbitrary and irregular, as if one were looking at neighboring peaks in a mountain chain. Each site is characterized by a resource fountain, an upwelling of the basic resources at that site. If we think of time as divided into discrete steps, as in a digital clock, then the fountain specifies the amount of each resource that appears at that site on each time-step. The amount varies from site to site and may range from 0 upward. One site may have no input of any resource, a "desert," while another may specialize in a high input of resource *b*, a "water spring," and still another may have a moderate input of all resources, a "pond." Agents interact at sites and a site can hold many agents.

MODEL 1: OFFENSE, DEFENSE, AND A RESERVOIR

In model 1, an agent has only two components: a *reservoir* for containing resources it has collected, and a single "*chromosome*" string, constructed of resource letters, that specifies its capabilities (see Figure 3.2). Let me emphasize that this so-called chromosome has only a few of the characteristics of a real chromosome. The terminology is suggestive, and there are similarities (more in later models than here), but real chromosomes stand in a *much* more complex relation to an organism's overall structure. Two critical characteristics are retained: (1) the chromosome is the agent's genetic material, and (2) the chromosome determines the agent's capabilities. In particular, in this model, an agent's ability to

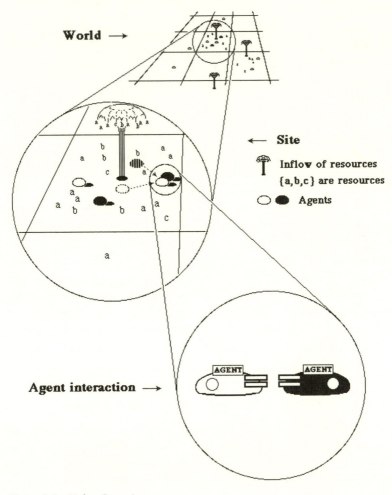

Figure 3.1   Echo Overview.

interact with other agents depends on tags specified by segments of the chromosome string. The mode of interaction is reminiscent of the way antibodies and antigens interact, although it can encompass a broad range of interactions of other real agents.

The crux of the Echo models is the requirement that an agent can reproduce only when it has acquired enough resources to make a copy of its chromosome string. The agent's fitness, its ability to produce offspring, is thus implicit in its ability to collect resources. Again, there are differences from real organisms. Here the chromosome stands in for all of the agent's structure, both cytoplasmic and nuclear. This representation buys a considerable simplification in the definition of structure and fitness. An agent can acquire resources either from the site it occupies or through interaction with other agents at the site.

In this first, simple model, each agent has a chromosome that does nothing other than specify two tags, an *offense* tag and a *defense* tag. All interactions in the model are mediated by these tags. When two agents encounter each other at a site, the offense tag of one agent is matched

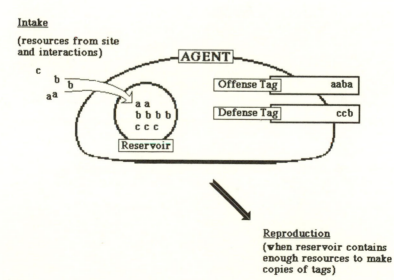

Figure 3.2   An Agent in Echo.

against the defense tag of the other agent, and vice versa. The object is to use the closeness of the matches to determine how resources are exchanged between the agents (see Figure 3.3). For example, if the offense tag of one agent is well matched to the defense tag of the other, it will acquire most of the other agent's resources, perhaps even resources tied up in its chromosome (thereby "killing" it). On the other hand, if the offense tag is poorly matched to the other's defense tag, the agent will receive only some surplus from the other's reservoir, or perhaps nothing at all.

To determine how well the offense string of one agent matches the

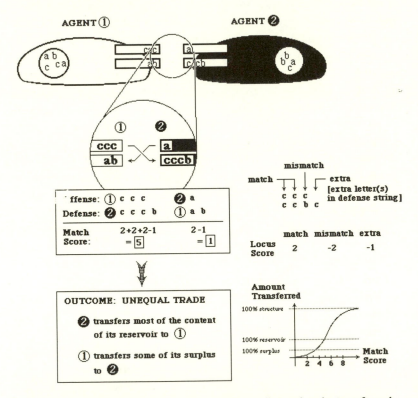

Note that a high match score causes resources (letters) to be transferred from the structure (tags) of the defendant, causing its demise.

Figure 3.3   Resource Exchange.

defense string of the other, the tag strings are first lined up so that their left ends are coincident. Then, a *match score* is determined by going down the strings position by position. At each position a value is assigned from a table that gives a value for each possible pair of letters (see the Locus Score line in Figure 3.3). For example, a *b* matched against a *b* might add 2 points, while a *b* matched against a *d* might subtract 2 points. If one tag is longer than the other, then each position without a paired letter counts for a fixed number of points (positive or negative). The overall match score is simply the sum obtained by adding these points.

In this model the possibilities for a given agent depend entirely on the pair of tags it carries. We can even extend this construct to interactions with the site itself, by assigning defense tags to the site. The agent acquires resources in proportion to the ability of its offense tag to match defense tags in other agents or sites. It avoids losses of resources in proportion to the ability of its defense tag to avoid matches with offense tags of other agents.

At first glance, it might seem that this version could be further simplified by allowing only one tag per agent. However, a bit more consideration shows that we would lose a vital property of *cas* interactions thereby. A single tag for each agent would force *transitivity* of interactions: if agent A can "eat" agent B and agent B can eat agent C, then with a single tag it would follow, under transitivity, that agent A can eat agent C. *Cas* interactions do *not* usually satisfy this property. In a real ecosystem hawks eat rabbits and rabbits eat grass, but hawks do *not* eat grass. The use of two tags allows us to avoid this constraint (see Figure 3.3).

Even this simple version of Echo offers interesting relationships between agents, once we set aside transitivity. For example, there is an interesting triangular relation, described by Hölldobler and Wilson (1990) in their monumental work, *The Ants*, that can be imitated in Echo (see Figure 3.4). One corner of the interaction triangle is occupied by a caterpillar that exudes a kind of nectar on its skin. Another corner is occupied by a fly that lays its eggs on the caterpillar, thereby becoming a predator through its larva. The third corner is occupied by a species of ant that is a ferocious predator on the fly. The ant is attracted to and consumes the caterpillar's nectar, but it is not a predator on the caterpillar. When the caterpillar is surrounded by ants it, of course, suffers much less predation by the fly. In effect, the caterpillar trades some of its resources for protection. This triangle is a stable relationship that changes drastically if one of the elements is removed.

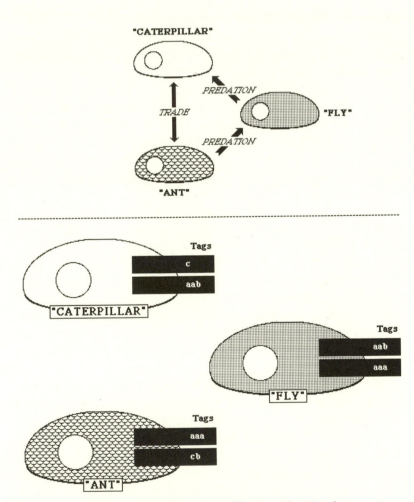

Figure 3.4 Echo Simulation of the Caterpillar-Ant-Fly Triangle.

This triangle provides an interesting test of Echo in several ways. First, there is an "existence" question: can we design tags for three different kinds of agents that allow trading between two of the agents, while retaining the predation relations among the three? The answer is yes (see Figure 3.4). Second, can we set Echo running with populations of these agents, and observe a persistent triangular relationship? The answer again is yes, though there are sometimes surprising developments over long periods of time. It is even possible for the top predator, the ant, to die out, leaving an oscillatory predator-prey relationship between the fly and the caterpillar—a relationship of the kind described by the Lotka-Volterra equations (Lotka, 1956). Finally, can we observe the evolution of such a triangle from a simpler starting point? At this point we don't know. The experiment has not yet been tried.

Extending the Basic Model

Although we can learn more from the basic model, that model is only a step toward modeling the complexities of a full-fledged *cas*. In particular, the basic model does not provide enough apparatus for a broad study of the way in which complex hierarchical structures emerge. Yet hierarchical structures are a pervasive feature of all *cas*. This section describes extensions that broaden Echo to the point where such phenomena can be examined.

In trying to model phenomena as broadly described as "complex hierarchical structures," we need to have one or more well-described examples in mind. The example that has guided much of my own work in this area is the embryogenesis of metazoans—the process whereby a fertilized egg progressively divides until it yields a mature many-celled organism that reproduces by producing another fertilized egg. The structure of a mature metazoan, such as a mammal, is incredibly complex, containing such complex hierarchical by-products as nerve networks, immune systems, eyes, and so on. An anatomist will tell you that such structures can really only be understood in terms of their origin and development in the maturing metazoan. And so it is with other *cas*. We can only understand one of these "patterns in time," be it New York

City or a tropical forest, if we can understand its origin and the way in which it has developed.

Just what happens as a fertilized egg develops into a complicated metazoan, say a tiger? A tiger has a hundred billion cells, more or less, organized in ways that make our most complicated computers look absurdly simple. Much of the development process is obscure even now, but we do have an outline of the main events. The process begins with the fertilized egg dividing into two cells, followed by further divisions that provide further doublings. These doublings cause a rapid increase in the number of cells (thirty doublings is enough to provide a billion cells). The offspring cells do not wander off as free-living entities; instead they adhere to their parent cells and to each other. Soon the number of cells increases sufficiently that there is a ball of cells with an interior and an exterior. The concentration of various metabolites—biochemical products of cell reactions—begins to vary from cell to cell. Some metabolites diffuse away from the exterior cells, while remaining in high concentration in the inner cells, and so on.

It is well known that changing concentrations of metabolites in a cell can cause different genes in the cell's chromosomes to be turned on and off. That is, the cell can respond to certain metabolites by starting up new activities while shutting down others. Cells with exactly the same chromosomes thus can have very different activities and forms. In a metazoan such as a tiger, this factor, more than any other, accounts for the immense differences among its constituent cells. A tiger's nerve cells are very different from its skin cells, even though both carry the same chromosomes. As the cells in the developing embryo increase in number, different genes turn on and off, causing even greater variation in the concentrations of metabolites in different cells. This change, in turn, alters the way the cells adhere to each other, giving rise to changes in the shape of cell aggregates. The initial ball of cells goes through an increasingly intricate set of transformations, eventually leading to local structures that become organs, networks, and the like.

My object, then, is to extend Echo so that it can mimic the process of producing a complex well-organized aggregate from a single "seed." Although the short précis just given does not do justice to the subtleties of the process of embryogenesis, it does suggest some mechanisms that Echo should include:

1. We need to add some means whereby agents can adhere to one another. It should include a provision for the formation of boundaries that enable the resulting aggregates to form functionally distinguished parts.

2. We need to enable an agent to transform resources, to mimic a cell's ability to transform abundant resources, at a cost, into needed resources in short supply.

3. We need to extend the definition of the chromosome string, so that parts of it can be turned on and off in a way that affects the interactions of the agents involved. Moreover, the process of turning parts on and off must be made sensitive to the activities of the agents, mimicking the effect of the metabolites in biological cells.

In adding capabilities to the Echo model, we want to retain the simple format of the agents in the basic model. In particular we want to retain three features: (1) the simple string-specified structure, (2) reproduction limited by resource acquisition (implicit fitness), and (3) interaction mediated by tags. The only way I can see to provide a chromosome with "switchable" genes, while retaining this format, is to treat the agents as organelles or compartments in a more complex, cell-like entity. That is, the agents, with their fixed structure, would be aggregated into a more complex variable structure, which I'll call a *multiagent*. With care, we can supply the multiagent with a chromosome that will be passed on to its offspring, while allowing the set of primitive agents (organelles) to vary from parent to offspring. That is, the multiagent chromosome describes the range of agents (organelles) the multiagent *can* contain, but the multiagent's offspring will contain only *some* of these agents. If we make the agents contained in the offspring dependent on activities within the parent multiagent, we get the effect of turning genes on and off. Then these cell-like multiagents can reproduce and aggregate into variegated, hierarchical structures that mimic metazoans. That, in brief, is the line we shall follow.

The simplest implementation I've been able to conceive within these constraints requires that the primitive agent be supplied with an additional five mechanisms, over and beyond the tag-mediated interaction and reproduction provided by the basic model:

1. A mechanism that allows selective interaction. An *interaction condition* checks a *tag* in the other agent to determine whether or not an interaction takes place (much as the condition in a rule checks a message).

2. A mechanism that permits resource transformation. An agent is provided the capability of transforming one resource into another, at the cost of gathering the resources necessary to define a *transformation segment* in its chromosome string. For example, with an appropriate transformation segment, an agent may transform an abundant resource into one it needs for reproduction. This process opens avenues for specialization of the agents in a multiagent.

3. A mechanism that determines adhesion between agents. This mechanism is implemented in terms of an *adhesion tag*. The amount of adhesion between two agents is determined by the degree of match between their adhesion tags.

4. A mechanism that allows selective mating. Implementation is by means of a *mating condition* that checks the interaction tag of a potential mate. A pair of agents having enough resources to reproduce will produce offspring by crossover if their mating conditions are mutually satisfied. This mechanism is not directly implied by the embryogenesis précis, but it makes the emergence of species possible.

5. A mechanism for conditional replication. A *replication condition* checks the activity of other agents that belong to the same multiagent aggregate. Even after an agent has collected enough resources to make a copy of its chromosome string, it only

reproduces if its replication condition is satisfied by the activity of some other agent in the multiagent. This mechanism is the one that has the effect of turning genes on and off.

In the next section, by adding one of these mechanisms at a time, I produce a sequence of increasingly sophisticated versions of Echo. As I add each mechanism, I use the syntax provided by Echo to redescribe the additional capabilities. If my conjectures are correct, the final model in the sequence should enable us to mimic the embryogenesis of multicellular organisms, or the origins of multiagent organizations such as Adam Smith's pin factory.

Each of these mechanisms is surprisingly easy to implement in a computer, though the verbal descriptions that follow are at times intricate. While the details do show that the mechanisms fit within the Echo framework, they do not enter much into the discussions that follow. If you, the reader, are willing to accept on faith the fit between the added mechanisms and Echo, then you can skip the next section, where the details are given, without substantially jeopardizing your ability to follow subsequent sections.

The Extensions

As promised, each model in this sequence extends the previous model by adding a single mechanism. The last model in the sequence implements the précis given above.

MODEL 2: CONDITIONAL EXCHANGE

The object now is to give each agent the possibility of rejecting exchanges with other agents. To accomplish this, we retain a single "chromosome" for the agent, but that chromosome is now divided into two parts, a *control* segment and a *tag* segment (see Figure 3.5). The control segment provides an *exchange condition* that checks the offense tag in the other interactant's chromosome; the exchange condition treats that tag much as a rule treats messages in a rule-based agent.

Because tags are defined over the resource alphabet, the exchange condition responds to strings over the resource alphabet, rather than to the binary strings used for messages in the rule-based system. To define the exchange condition, we use a "don't care" symbol, as in Chapter 2. We can avoid adding a new symbol to the resource alphabet by simply designating one of the symbols already in the alphabet as the don't care symbol. That is, in our earlier example using the alphabet $\{a,b,c,d\}$, we would restrict the definition of tags to the subalphabet $\{a,b,c\}$, constructing strings over the full alphabet $\{a,b,c, \#(=d)\}$ to define conditions.

Tags may be of different lengths, unlike the standardized length of messages, so let's alter the definition of a condition accordingly. To accommodate arbitrary lengths, we treat the last specified letter in the condition string as if it were followed by an indefinite number of don't

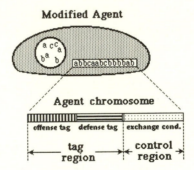

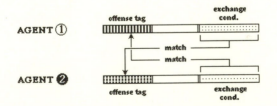

Figure 3.5   Agent Chromosome with Added Exchange Condition.

care symbols. That is, the condition $b \# b$ $(= bdb)$ is treated as if it were the condition $b \# b \# \# \# \#$ .... Here are a couple of examples: The condition a accepts for resource exchange any agent having an offense tag that starts with an a. That is, it accepts any offense tag from the set {a, aa, ab, ac, aaa, aab, aac, aba, abb, . . .}. Similarly, the condition bcb accepts any offense tag that starts with bcb. The condition $b \# b$ is a bit more complicated, accepting any offense tag that has a b at the first and third positions, namely, the set {bab, bbb, bcb, $baba$, $babb$, $babc$, $babaa$, . . .}.

The condition is used as follows. When two agents encounter each other, the exchange condition of each agent is first checked against the other agent's offense tag. If the conditions of both agents are satisfied, then the exchange takes place. If neither condition is satisfied, then the interaction is aborted. If the condition of one agent is satisfied but not the other, then the agent with the unsatisfied condition has a chance of "fleeing" the interaction. In the simplest case, it does so by aborting the interaction with some fixed probability.

MODEL 3: RESOURCE TRANSFORMATION

The ability of cells or factories to transform resources into new forms is a valuable property worth capturing in Echo. As we will see, this option can be critical for certain agents if a particular resource is in short supply. In particular, when we get to layered multiagents, resource transformation offers interesting opportunities for specialization. Again, I'll take the simplest possible approach, leaving elaborations for future models.

Consider the "renewable" resources that underpin the agent structures in Echo. We can think of each of these resources as a molecule having an interior structure. Using cellular biology as a guide, we can think of transforming one resource into another by rearranging the "molecular" structure. In a biological cell such transformations are controlled by enzymes (the potent biological catalysts that can speed a reaction by a factor of 10,000 or more). Our object is to provide agents with counterparts of enzymes.

Because I am trying to avoid questions concerning the metabolism of assembly, I prefer not to become concerned with the details of resource structure. Rather, my objective is to provide agents with a direct way of transforming resource letters, {a,b,c,d} in our running example, into other resource letters. The simplest way to do so is to add a subsegment to the chromosome for each transformation desired. It is important that there be a "cost" to this operation; otherwise, resources would be freely interchangeable, and we would have no way to study the effects of shortages or resource bottlenecks. The cost, as in earlier models, will be a requirement that agents use resource letters to build the enzyme subsegment specification. For each transformation there must be an enzyme subsegment of the control segment, and the cost is the effort required to collect the additional letters needed to specify these transformation subsegments.

The transformation subsegment must, at a minimum, specify the letter to be transformed and the letter that will result from the transformation (see Figure 3.6). The simplest designation would use just the two letters involved. If a is to be transformed into b, then the transformation subsegment would be the substring ab. If the transformation is to be made more costly, then additional letters are required, so that, for example, the transformation segment for the transformation of a to b would be the substring $abcccc$. We can think of the a and b in this substring as specifying the "active sites" of the enzyme, and the $cccc$ as specifying the structural part of the enzyme, the part that places the active sites in a proper three-dimensional configuration.

There is still the matter of the "rate" of the transformation invoked by a transformation subsegment. How much a will be transformed into

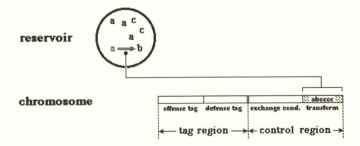

Figure 3.6   Resource Transformation.

b if the *ab* subsegment is present? It seems reasonable to confine the transformation to resources the agent has collected in its reservoir. That is, the transformation can only take place if there are copies of the letter *a* in the agent's reservoir. A transformation will pay off if (1) the definition of the agent's chromosome requires several copies of a target letter that is in short supply, and (2) the rate of transformation is fast enough that several copies of the resource letter can be transformed during the agent's life span. Otherwise, the investment of resources to define the transformation subsegment can never "pay off." For instance, it takes one instance of the letter *b* just to define the *ab* transformation subsegment, so the investment cannot under any circumstances pay unless at least two copies of the letter *b* can be obtained by transforming *a* into *b* during the agent's life span. Because the shortest life span is one time-step, let's set the rate at two letters per time-step. Then even short-lived agents can benefit from a transformation subsegment.

It seems natural to adopt the convention that multiple copies of the transformation segment multiply the transformation rate. If an agent has two copies of the *a* to *b* transformation segment in its chromosome, it will transform four copies of *a* into *b* per time-step, given four or more copies of *a* in its reservoir. It will pay to have multiple copies of the transformation segment if the target letter *b* is in short supply, the letter *a* is regularly in surplus, and the agent uses *b* extensively in its chromosome.

Clearly, we are free to choose different transformation rates in different models, and we can even choose different rates for different letters in the same model. The relation between these transformation rates and the site input rates for the basic resources will certainly affect the evolution of the model. Evolution, by working on the transformations, should "flatten" differences caused by different site input rates.

MODEL 4: ADHESION

Adhesion provides a way of forming multiagent aggregates. These aggregates are reminiscent of colonial organisms (sponges and jellyfish) and metazoan organisms (plants and animals). Agents selectively adhere to each other and even form "layers." As a result, they move and interact as units. Individual agents in the aggregate can adapt, over successive generations, to take advantage of the specific environment provided by the other agents in the aggregate. One agent in the aggregate might specialize for offense or defense, while a second might specialize in resource acquisition. If these two kinds of agents exchange appropriate resources, then the aggregate and the agents therein will collect and protect resources more efficiently, and therefore reproduce more rapidly.

It is as if the ants in our caterpillar-ant-fly triangle were permanently attached to the caterpillars, instead of being independently mobile. The caterpillars can reduce to a minimum the resources committed to offense tags, while the ants can specialize their tags to efficient offense without concern for resource acquisition.

Once aggregates start to form and survive, interactions and exchanges can evolve into ever more sophisticated configurations. One kind of agent, by collecting and supplying a particular resource, can induce a second kind of agent to specialize by taking advantage of an assured supply of that resource. Some kinds of agents may also gain a competence for resisting such inducements. The interplay of *induction* and *competence* is a major aspect of developmental biology (see, for example, Buss, 1987).

How can we implement conditional adhesion in Echo? Once again tags, and the matching of tags, will play a key role. The procedure will be much like the procedure for resource exchange. When agents come into contact they will be checked for adhesion, as in the Chapter 1 example of the sticky billiard balls. To implement this operation, a new tag that mediates adhesion is added to the tag segment of the chromosome. We can think of this tag as a kind of *cell adhesion molecule* (see Edelman, 1988).

The interaction proceeds as follows. A pair of agents is selected for interaction as in resource exchange. For adhesion it is often useful to pair a parent with its offspring; this coupling facilitates an aggregate that grows from a single agent, much like the growth of a metazoan organism from a fertilized egg. It is important to allow agents of the same kind, as is often the case for parent and offspring, to have less than perfect adhesion. To accomplish this, the adhesion tag is *not* matched to the adhesion tag on the other chromosome; if this were done, agents of the same kind would always match perfectly, producing maximal adhesion. Instead, the adhesion tag of each agent is matched to the *offense tag* on the chromosome of the other agent (see Figure 3.7).

Match scores are then calculated. If each agent has a score close to zero, then no adhesion takes place between the two agents. If at least one of two match scores is not close to zero, then adhesion does take place. The configuration induced by the adhesion depends on an additional mechanism, boundary formation.

Boundaries

Boundaries provide a simple way of aggregating agents into layers somewhat like those of an onion, and they are used to constrain agent interactions. Each agent, at the time of its formation, is assigned to exactly one boundary. Even an isolated agent that adheres to no other agents is assigned to a unique boundary that contains that single agent.

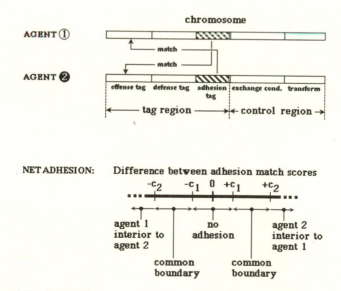

Figure 3.7　Agent Chromosome with Added Adhesion Tag.

However, a boundary can contain many agents. The simplest nontrivial aggregate is an aggregate that has only one boundary, with all agents in the aggregate belonging to that boundary.

It is useful to array boundaries into configurations a bit more complicated than simple layering. Rather than constraining each boundary to contain a single interior boundary, as in the case of the onion, we allow a boundary to contain *several* boundaries at the next level inward, like an egg with multiple yolks. The simplest example of this configuration is an outer boundary that contains two interior boundaries side by side (see Figure 3.8). We can describe the progres-

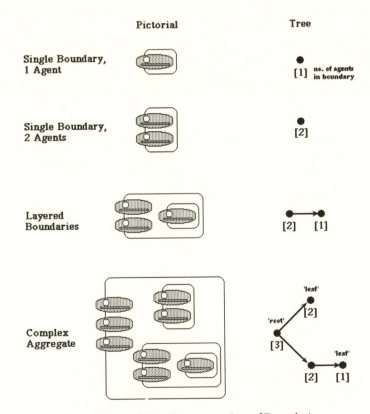

Figure 3.8   Boundaries and Tree Representation of Boundaries.

sive, possibly multiple, inclusions by using a kind of family tree. The outermost boundary is represented by a node at the root of the tree. Each of the boundaries directly included within the outermost boundary is represented by a node connected to the root. An included boundary can, in turn, contain further boundaries. A new node is added for each "deeper," second-level boundary, and it is connected to the node representing the boundary containing it. This process is repeated until we reach the innermost boundaries. Those are represented by nodes that constitute the "leaves" of the tree (no further connections).

Boundaries constrain agent interactions as follows. An agent can only interact with agents belonging to the same boundary, or with agents belonging to adjacent boundaries. A boundary is adjacent to a given boundary if it is directly exterior to (toward the root of the tree), or directly interior to (toward the leaves of the tree), or resides alongside (at the same level as, hence directly connected to the same node as) the given boundary (see Figure 3.9). The set of agents with which a given agent can interact is called its *domain of interaction*. It is convenient to think of the site itself, with its supply of renewable resources, as a boundary exterior to all the agents the site contains. Only agents on the outermost boundary of an aggregate have a domain of interaction that includes other aggregates at the site. This domain of interaction includes solitary single-agent aggregates, as well as the renewable resources offered by the site.

The boundary to which an agent belongs is decided, via the adhesion match scores, at the time it is formed from its parent. Generally, each newly produced offspring undergoes an adhesion interaction with its parent, but it also is useful to give the offspring a kind of mobility, so that adhesion sometimes involves an agent other than the parent. To simulate this mobility, another agent is sometimes selected at random from within the parent's domain of interaction; this choice occurs with a probability that is a fixed parameter of the model. Match scores are calculated for the pair consisting of the newly formed offspring and the parent or selected agent, and the outcome is determined as follows:

1. If both match scores are low, then (as mentioned earlier) the agents do not adhere. If the parent belongs to an aggregate, the offspring is ejected from the aggregate and becomes a new one-boundary, one-agent aggregate. This ejected offspring, if it has an appropriate structure, can become the seed of a new aggregate similar to the one containing the parent.

2. If the two match scores are close to each other in value and not close to zero, the offspring is placed in the boundary of the selected agent.

3. If the match score of the selected agent is substantially higher than that of the offspring, the offspring is placed in the bound-

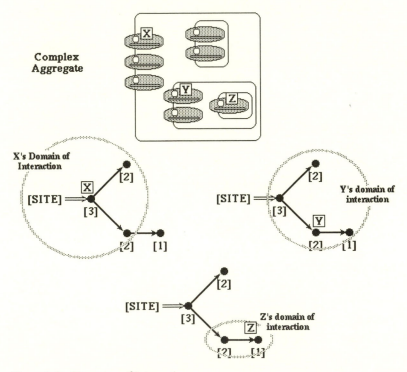

Figure 3.9   Domains of Interaction.

ary immediately interior to the selected agent's boundary. If the parent's boundary has no interior boundary, then one is formed to contain the offspring; this way an aggregate develops additional boundaries as its agents reproduce. The result is a kind of developmental induction on the part of the parent, where the offspring is forced to occupy a position it might not otherwise occupy.

4. If the net score is high negative, then the effect is reversed; the parent is forced to the interior of the boundary it occupies.

Options and Tests

If desired, adhesion interactions can take place at times other than the formation of offspring. Under such an arrangement the interactions can occur on a "random contact" basis, as in the exchange interactions. Agents in the same domain of interaction are paired, as for resource exchange, and the scoring scheme just described is used to determine the outcome. With this provision an aggregate changes at a rate determined by the frequency of the adhesion pairings. Adhesions already in place may be changed by these interactions. For instance, free agents could collect to form an aggregate, somewhat in the way the amoeboid individual cells of slime mold aggregate to form a stalk-like aggregate (a surprising sequence nicely described in Bonner, 1988). Or an agent in an aggregate may be expelled as a free agent, to become a seed for a new aggregate, if it has an appropriate chromosome.

Possible effects of conditional adhesion can be tested by setting up designed aggregates in Echo (one could set up an aggregation imitating Adam Smith's pin factory). As with the caterpillar-ant-fly triangle, the aggregate is tested for stability and for its ability to reproduce under the laws of Echo. A more severe test, and a more interesting one, would be to see if free agents can aggregate to become more efficient at collecting and processing some resource. Such a study would move us in the direction of understanding how Adam Smith's pin factory first originated from an aggregation of individual craftsmen.

MODEL 5: SELECTIVE MATING

Selective mating provides a way for agents to choose among potential mates, so that crossover occurs only with selected kinds of individuals—the origin of species within Echo. As with resource exchange and conditional adhesion, this interaction is tag mediated.

Selective mating is implemented by adding a *mating condition* to the control segment of the chromosome (see Figure 3.10). This condition is specified in the same way as the exchange condition, and it is matched against the already extant offense tag of the potential mate. (We could, of course, provide a completely new tag for this purpose. But it seems that many of the effects of selective mating can be attained without adding another tag to the chromosome.)

Selective mating is initiated once an agent has collected enough resources to make a copy of itself. It then initiates a search for a mate with which it can exchange chromosomal material. There are many ways to do this, one of the simpler of which is to randomly select the potential mate from the set of agents that are (1) ready to reproduce, and (2) within the domain of interaction of the given agent. If the tag-mediated selective mating conditions of *both* agents are satisfied, mating proceeds. Copies of the parents' chromosomes are made, using the resources in their reservoirs. The copied chromosomes are crossed, mutations take place, and the two resulting offspring are added to the population at the site. This procedure is a bit like conjugation between different mating types of paramecia (a process described in any standard genetics text such as Srb et al., 1965). If one or both of the mating conditions are not satisfied, the mating is aborted.

Note that an agent may be more or less selective concerning the

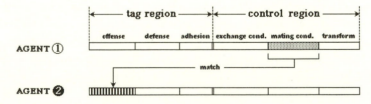

Figure 3.10   Agent Chromosome with Added Mating Condition.

agent it will accept as a mate, depending on the specificity of the mating condition. Some agents may accept almost any other agent, while others may be quite selective. This distinction gives considerable scope to the evolutionary processes in Echo. It will be interesting to see what environmental conditions favor the tight mating criteria typical of mammals, and to contrast these with environmental conditions favoring the more relaxed criteria typical of plants.

There is still one problem that must be resolved in implementing selective mating. We want to study complex adaptive systems where there are limitations on the number of agents a site can sustain. Earlier, when we were dealing only with free agents, we did so by having the offspring replace an agent drawn at random from the site, thereby imposing a death rate that balanced the birth rate. This procedure makes less sense now that agents, because of adhesion, have locations within an aggregate. When a new agent is formed within an aggregate, which agent, if any, should be deleted? There are many options, but a simple one is to set a random death rate for all agents, decoupling death from birth. That is, all agents have an average life span, and agents are removed from their boundaries whenever chance, determined by the random death rate, decrees. Subsequent replacement is indirect. Each offspring formed is immediately tested for adhesion and is placed in the boundary so determined. The offspring is immediately added to the boundary, without replacing any agents there. Only the overall random death rate will eventually balance the process.

MODEL 6: CONDITIONAL REPLICATION

With conditional replication we can, finally, construct a simple model of metazoan embryogenesis within the Echo framework. Metazoans accomplish the quite remarkable feat of developing from a single cell, a fertilized egg, into a multicelled organism with a great diversity of cell types. Yet all the cell types within this organism (with a few exceptions, such as germ cells and some cells in the immune system) contain the same chromosomes. How is this possible?

It is not just this question that impels me to add morphogenetic processes to Echo. All *cas* exhibit phases of increasing organization as they evolve, but we have little that connects *cas* mechanisms to this increasing organization. The dynamic in most *cas* is so intricate that it beggars standard scientific techniques for treating dynamics. The mathematical models we have simply do not encompass the dynamics of morphogenetic processes, and controlled experiments with the systems themselves are difficult or impossible.

One of the difficulties centers on the symmetry breaking that goes on in these morphogenetic processes. A metazoan grows from a single fertilized cell via successive generations of cell division. However, this cluster of cells soon loses its spherical symmetry, for it goes through a series of stages where physical symmetries are lost, one after another. And this is only the outward appearance. We know that the chemical constitution of these cells becomes progressively more diverse, breaking even more symmetries. It is difficult to treat such processes with partial differential equations (PDE's), our traditional mathematical tool for understanding dynamic processes.

Turing (1952) did manage to use PDE's to design a model that started from symmetric initial conditions, but produced an asymmetric variegated pattern, much like the color pattern of a Holstein cow. Even this simple formulation was mathematically intractable: Turing could observe specific examples of the dynamics, but he could derive no general consequences from the mathematical model. In fact, he depended on a computer-based version of the model to exhibit the dynamics of asymmetric pattern formation. Little has been done mathematically since then, and the problem remains much as it was.

As an aside, I note that part of the overall difficulty that attends attempts to model morphogenesis is inadvertent and unnecessary. From training, habit, and previous success, physicists and mathematicians usually describe dynamic processes in terms of PDE's. Maxwell's nineteenth-century description of electromagnetic dynamics and Einstein's twentieth-century theory of relativity both use simple, beautifully symmetric sets of PDE's. Those two triumphs of theoretical physics underpin most present-day technology. The advent of the computer did little to change this approach. Models of dynamic processes are first written in (continuous) PDE's, then these equations are translated to (discrete) computational routines. However, this labored approach is *not* necessary. Models can be directly written in terms of conditional actions, as in our description of adaptive agents, and other combinatorial operations such as crossover. These conditional/combinatorial operations are only awkwardly captured by PDE's, so a direct approach substantially enlarges the scope of rigorous modeling.

My own view is that a move toward computer-based models that are directly described, rather than PDE-derived, will give handsome returns in the study of *cas*. I do not think we will understand morphogenesis, or the emergence of organizations like Adam Smith's pin factory, or the richness of interactions in a tropical forest, without the help of such models. Our experience to this point with direct models suggests that they can exhibit the combinatorial complexities of developmental processes. If this is true, such models offer the possibility of controlled experiments that can suggest both guidelines for examining real organizational processes and guidelines for mathematical abstractions of organizational dynamics.

In building direct computer-based models of morphogenesis, we can be guided by the now-extensive knowledge of the mechanisms employed by metazoans in morphogenesis. This knowledge, hard won by molecular geneticists, involves intricate pathways; but there is a simple statement that summarizes the basic idea. Metazoans exhibit increasing organization and diversity as they develop because the genes in their chromosomes can be turned on and off (there is a good discussion in the text of Srb et al., 1965, in the section titled "The Modulation of Gene Action"). To give a little more detail, genes that are on are expressed by the cell's construction of the enzymes they encode. Enzymes are such effective catalysts that they redirect the reactions in the cell. When different genes are on, different enzymes and different reactions result, leading to different structures. As a result, a single organism has cells as different as nerve cells, muscle cells, and blood cells—even though all the cells have the same chromosomes.

This outlook takes us part of the way, but it leaves us with a further question. How are the genes turned on and off? Again, molecular genetics has something to tell us. Strings of genes in a chromosome often have "headers"—tags again—that are sensitive to the biomolecules present in the cell (see Srb et al., 1965). If one of these molecules attaches to the header, it can interfere with the expression of genes downstream from the header. The genes are *repressed* (turned off). Other molecules can clear the header, *derepressing* the genes (turning them on).

The genes themselves can, through the enzymes, favor or disfavor the production of a wide variety of biomolecules. This fact opens the possibility of intricate feedbacks whereby one gene, through its bio-molecular by-product(s), can turn other sets of genes on or off. In effect, the chromosome encodes a computer program with all sorts of conditionals. Perhaps we can directly construct a relatively simple computer-based model, if we can set aside some of the metabolic details without losing the essence of the process.

Multiagents and Agent-Compartments

With these guidelines the question concerning mechanisms for morphogenesis becomes: How can we imitate the repression and de-repression of genes within Echo's limited format? So far we have attempted to keep the individual agents quite simple, so the chromosome of a given agent does *not* offer an array of "genes" (conditions and specifications for tags) that can be turned on and off. In biological terms the agents come closer to representing the organelles in a cell, with their fixed functions, rather than the flexible organization of a whole cell.

We need to try to aggregate the simple agents into something that comes closer to a whole cell, with its multiple functions. This coming together is reminiscent of Margulis' theory of the origin of eu-karyotes, the advanced cells that give rise to metazoans (see, for example, Sagan and Margulis, 1988). According to this theory, an eukaryote is a symbiotic amalgam of simpler, originally free-living, precursor cells. The amalgam is formed when one precursor engulfs another but fails to digest it. An aggregate at this level, call it a *multiagent* (short for multicompartment agent), would have its structure determined by a chromosome that amounts to a concatenation of the chromosomes of the component agents (see Figure 3.11). If properly done, the multiagent would accumulate an array of genes that could be turned on and off. The multiagents could then further aggregate, playing the role of cells in a metazoan.

Following this line, I will retain the agents so far defined as the primitives of the system. They will serve as organelles or compartments in the multiagent. To emphasize this aspect, I'll call the primitive agents *agent-compartments*. We have to distinguish carefully between the multiagent's chromosome and the compartments that chromosome describes. On the one hand, we want to define the multiagent's chromosome as the concatenation of the chromosomes of its component agent-compartments. On the other hand, we want successive generations of the multiagent's offspring to have different arrays of agent-compartments (so that the multiagent can carry out different functions). But then the multiagent's chromosome must not depend directly on the agent-compartments present within it; otherwise the multiagent's chromosome would change from one generation to the next as its compartment-agents changed. The multiagent can retain

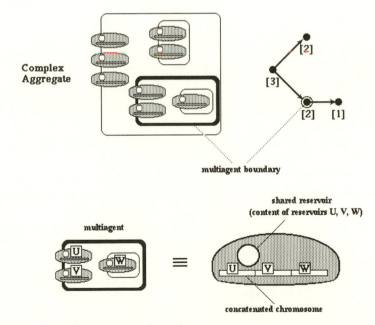

Figure 3.11 Characteristics of a Multiagent.

hard-won adaptations from one generation to the next only if its chromosome remains invariant under these changes.

To resolve this quandary, we have to designate an initial or basic form for the multiagent, an ur-form, much like the fertilized egg from which the rest of a metazoan develops. This ur-form will have a chromosome that describes the full range of agent-compartments that the multiagent may exhibit under various conditions, and that chromosome will be carried from generation to generation.

Conditional Replication of Agent-Compartments

Our objective, then, is to design an aggregation procedure that (1) acts as a single chromosome for the multiagent, and (2) allows different parts of this chromosome to be active in different versions of the multiagents. The guiding biological analogy can carry us a bit further. It suggests that we think of a given agent-compartment as producing a key biochemical

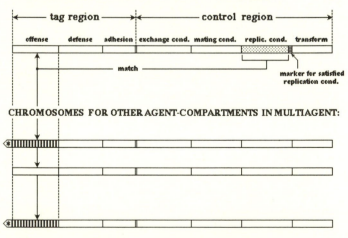

CHROMOSOME FOR REPLICATING AGENT-COMPARTMENT:

⟨✳⟩ = active agent-compartment: an agent-compartment becomes *active* when it participates in an interaction (exchange or adhesion).

Figure 3.12   Agent Chromosome with Added Replication Condition.

when it enters into an interaction. Let's call such an agent-compartment *active.*

We can implement this suggestion by setting up conditions, similar to the headers mentioned earlier, that make the replication of an agent-compartment dependent on the activity of other agent-compartments in the multiagent. That is, we replace the < biochemical / gene derepression / enzyme / new biochemical > sequence by an < active agent-compartment / condition / new active agent-compartment > sequence. Under this setup the replication of an agent-compartment is determined by a replication condition located in the control segment of the part of the multiagent's chromosome that specifies the agent-compartment. The agent-compartment can replicate only if that condition is satisfied by the activity of some other agent-compartment in the multiagent. In this way a multiagent can have an offspring multiagent in which some compartments are missing because their replication conditions were not satisfied (the corresponding genes were repressed). Note that the offspring multiagent's chromosome is unchanged, even though the set of compartments is different. Because the offspring multiagent can have a different array of agent-compartments from its parent, it can have different interaction capabilities, thus the multiagent mimics the flexibility of a metazoan cell.

Specifically, this process comes down to adding a *replication condition* to the control region of each agent-compartment (see Figure 3.12). This condition looks to the offense tags of the other active agent-compartments in the multiagent. The replication condition is satisfied only if at least one active agent-compartment in the multiagent has an offense tag that meets the condition's requirements (see Figure 3.13).

At the time the multiagent replicates, each agent-compartment replication condition that is satisfied is marked. That is, each replication condition has an added marker bit which is set to 1 ("marked") if that condition is satisfied at replication time; otherwise it is set to 0 ("not marked"). Agent-compartments with (replication condition) markers set to 1 are considered to be "present" in the offspring; those with markers set to 0 are considered to be "absent," even though coded for in the chromosome (see Figure 3.13). An offspring multiagent can then differ from its parent in the number of marked conditions, even though it and its parent have the same (concatenated) chromosome. Only agent-compartments with marked replication conditions ("present") enter into interactions.

Multiagent Interaction

Finally, we have to be more specific about the relation between agent-compartment capabilities for interaction and multiagent capabilities for interaction. For example, what determines a multiagent's adhesion capabilities?

Here I invoke a simple principle that uses agent-compartment capabilities directly: all interactions between multiagents are mediated by

Example:

Replication Condition of Agent	Is Satisfied by Offense Tag(s) of Agent(s)
U	U, V
V	W
W	V

For instance activity of either agent-compartment U or V assures that agent-compartment U appears in the next offspring of the multiagent.

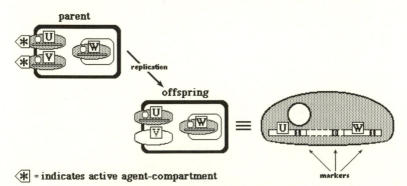

❖ = indicates active agent-compartment

Figure 3.13   Conditional Replication of a Multiagent.

their marked agent-compartments. It is easy to implement this principle if we follow our earlier approach for individual agents. There we selected two agents at random for each interaction. Now we select two aggregates in place of individual agents. In effect, aggregates move about the site as units. If one (or both) of these aggregates is a multiagent, we must determine the form and outcome of the interaction. To do this, we go one step further, randomly selecting one of the agent-compartments in the multiagent's outermost boundary (see Figure 3.14). Only agent-compartments having markers set to 1 are eligible for selection. The selected agent-compartments serve as the "point of contact" for the given multiagent interaction. A new selection is made each time multiagents come into contact. Once the point-of-contact agent-compartments have been selected, the interaction is carried out as described for individual agents in the previous models.

Interactions within a site center on the multiagents, but the details of the interactions still depend on the point-of-contact agent-compartments. Accordingly, the possibilities for interaction remain those described in the previous five models. The agent-compartments are still the primitives that mediate adhesion and the accumulation of resources.

The accumulation of resources within the reservoirs of the component agent-compartments brings up an additional question: how are the resources in these reservoirs distributed for reproduction of the multiagent? Several conventions could be followed here, but one seems particularly interesting. It treats a multiagent as an organization with shared resources (see Figure 3.11). With this convention, the contents

Once multiagents are paired for interaction, an agent-compartment in the external boundary of each multiagent is randomly selected as the "point of contact." The selected agent-compartments undergo an agent-to-agent interaction.

Figure 3.14   Multiagent Interaction.

of the individual agent-compartment reservoirs are available for reproduction of all parts of the multiagent chromosome, in contrast with
using the content of each agent-compartment reservoir only for reproduction of the part of the chromosome that describes that agent-
compartment. This convention allows a wide variety of specializations,
akin to the permanently joined caterpillar-ant discussed in model 4. For
example, one agent-compartment might specialize in accumulating, or
producing, resource b, even though it uses few b's in its own (portion of
the) chromosome. Under the shared reservoir convention, many paths
lead to enhanced reproduction rates, encouraging continued diver-
sification of the multiagents.

As with agents in the earlier models, multiagents continually
interact—even the multiagents within a larger aggregate. Each interaction typically changes the content of the reservoirs of the agent-
compartments involved. Because of the sharing, a multiagent's pos-
bilities for reproduction are modified. As in earlier models, the
ultiagent reproduces when it has enough resources in the reservoirs of
its agent-compartments to make a copy of its chromosome.

Distinguishing Multiagents from Other Aggregates

One last question about multiagents remains: When a multiagent resides within a larger aggregate, how do we distinguish it from the rest of
the aggregate? This distinction must be made in order to determine
which agent reservoirs are shared in reproduction. A closer look at
the organization of boundaries within an aggregate gives us a direct
approach. Obviously a multiagent, being an aggregate of agent-
compartments, must have an outermost boundary. So the question
becomes; How do we mark the boundary of an aggregation of agent-
compartments as the boundary of a multiagent? Once we make this
provision, we can define the chromosome of the multiagent and we can
provide for further layering and boundaries involving multiagents.

In thinking about ways to mark a multiagent's boundary, we must
also think about how that marking can originate and evolve. It is helpful
to return to the convention that an independent single agent is treated
as one-agent/one-boundary aggregate. Within this convention, we

might as well treat an independent single agent as a one-agent/one-boundary *multi*agent. That is, we treat an independent agent as the simplest multiagent. We can then think of starting Echo with only the simplest multiagents (the independent single agents), leaving it to evolution to provide more complicated versions.

Of the many possibilities for increasing the complexity of multiagents, one of the simpler ones is the following. Occasionally, "promote" an aggregate of the simplest multiagents to the status of a single multiagent, demoting the components to agent-compartments. To implement this idea, add a *multiagent boundary* marker bit to the boundary specifications. When the marker is 1 ("on"), the boundary is the boundary of a multiagent; otherwise, the boundary plays its usual role (see Figure 3.11). Then, when a pair of multiagents adhere to each other, we occasionally carry out the promotion/demotion procedure. That is, the marker for the boundary that contains the two multiagents is set to 1, and the markers for the boundaries of the two multiagents are set to 0 (see Figure 3.15). The result is a kind of mutation that produces a larger multiagent composed of the original pair of multiagents. Some care must be exercised so that the multiagent will not contain other multiagents. It is easy to invoke this constraint at the time the promotion/demotion procedure is executed.

We now have a way that complex multiagents can evolve in Echo, and we need only tidy up one detail concerning the multiagent's chromosome. The whole objective of adding multiagents to Echo is to facilitate the common-chromosome/variable-structure feature of metazoan cells. We know that we derive the multiagent's chromosome by concatenating the chromosomes of the component agent-compartments. In Echo we literally string the agent-compartment chromosomes together to form one long chromosome. This simple convention is the reason that we do not want a multiagent to contain other multiagents—the concatenation convention would become ambiguous. The multiagent reproduces when it accumulates enough resources in the reservoirs of its agent-compartments to copy the long chromosome. It is this chromosome that undergoes crossover and mutation and is then passed on to the offspring multiagent.

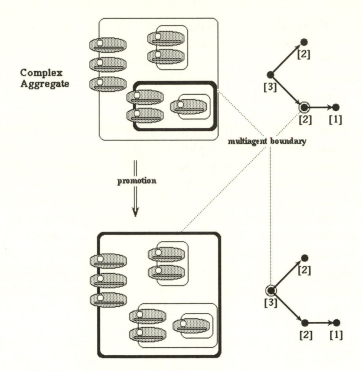

Figure 3.15   From Aggregate to Multiagent.

Summarizing

There certainly are other mechanisms that could be added to model 6, and there are modifications that could be made to the steps leading to this model, but model 6 gives a fair indication of the scope and intent of the Echo models. Let me summarize.

- Echo has a geography represented by a network of *sites*. Each site contains *resources* and *agents*.
- The resources are represented as a set of letters {$a,b,c,d, \ldots$}. Each site may have an upwelling or fountain that provides a

selection of resources on each time-step, though some or most sites may be barren. In effect, the resources are renewable.

■ The agents, called agent-compartments in model 6, have structures represented by stringing resource letters together. The strings are called *chromosomes*. (Again, I emphasize that these chromosomes are far removed, in both complexity and function, from biological chromosomes, though there are *some* similarities.) In addition, each agent has a reservoir for storing resources acquired through interactions with the site and other agents at the site. An agent has no other parts. In order to reproduce, an agent must collect enough resources through interactions to make a copy of its chromosome.

■ The chromosome of an agent in model 6 consists of a *tag segment* and a *control segment*. This chromosome provides the agent with three tags, three interaction conditions, a capability for resource modification, and a means of making an agent active or inactive. (I have tried to reduce this array, but so far have found no way to do so and still allow the scope and examples I have in mind.)

■ The tag segment contains three tags, an *offense* tag, a *defense* tag, and an *adhesion* tag. When two agents interact, the offense tag of each agent is matched to the defense tag of the other to determine the amount of resource exchange between the two (as in model 1); the offense tag is also used to constrain the conditional exchange, mate selection, and conditional replication interactions (models 2, 5, and 6). The adhesion tag determines the degree of adherence between two interacting agents (model 4).

1. The adhesion tag has some accompanying apparatus that plays a major role in the formation and evolution of organizations within Echo. When agents aggregate, they form extra-agent structures called *boundaries*. A treelike structure records boundaries and hence the

relative position of each component agent within an aggregate (model 4).

2. Sometimes aggregation results in a particular structure called a *multiagent*. Such a unit treats the chromosome of its component agents as a single chromosome, and it shares all the resources in their reservoirs for purposes of replication of the whole. Nodes in the tree structure that represent the boundary of a multiagent are marked accordingly (model 4).

■ The control segment contains three kinds of objects: conditions, resource transformations, and an activity marker.

1. There are three conditions: an exchange condition, a mating condition, and a replication condition (models 2, 5, and 6, respectively). Whenever agents are paired for interaction, each condition checks the *offense* tag of the other interactant's chromosome to determine whether or not the interaction will proceed.

2. There can be any number of resource transformations. Each designates a source resource and a target resource; when source is available in the reservoir, the resource transformation transforms it to the target at a fixed rate (model 3).

3. There is one marker in the control segment. If the marker is set to 1, then the multiagent uses the agent's tags to mediate its interactions; otherwise the multiagent acts as if the agent were not present in its aggregate (model 6).

What Has Been Left Out?

Echo is kind of a caricature because I have kept the mechanisms few and quite primitive. My bias is that simplicity, and elegance if you will, help us to describe complexities, as they do in mathematics. Equally important, keeping the mechanisms primitive helps us avoid "unwrapping," the bête noir of computer-based investigations of complexity. Unwrapping occurs when the "solution" is explicitly built into the program from the start. Consider a program that is supposed to discover a simple description of the movements of "the wanderers" (the planets) by using a compilation of their successive positions in the night sky (à la Kepler using Tycho Brahe's data—see Lodge, 1950). If the program is *explicitly* given ellipses centered on the sun as one of a few possibilities, we will learn little. We will have jumped over the complex reasoning that led from the wanderers' two-dimensional, S-shaped movements in the night sky to planets moving in three-dimensional space on sun-centered elliptic orbits. With unwrapping, the simulation reveals little that is new or unexpected.

Given this deliberate attempt at caricature, it is important to know what has been left out of Echo. In this respect, understanding Echo is not so different from understanding the relevance of a good political cartoon. We have to know what has been emphasized (or exaggerated) to make a point, and what has been left out as distracting from that point. Echo's design uses three major shortcuts:

- Details of metabolism, and assembly of resources into the agent's structure, have been omitted. Once the resources are acquired, they are automatically assembled into the required structure—the chromosome string—with no attempt to simulate the chemistry involved. (By progressively adding resource transformation capabilities to agents, the evolution of metabolism can be modeled with increasing verisimilitude.)

- The agent's internal structure—the phenotypic detail—is represented on the string that provides the agent's genetic legacy—the genotype. The agent does have a phenotype because it exhibits tags, and it conditions its interactions on the tags presented by other agents. In a biological cell these phenotypic characteristics would be biomolecules attached to organelles that are generated by decoding the genes. In Echo, however, these characteristics are presented on a string that plays the role of both the organelles and the chromosome that specifies them. (It would not be difficult to separate these functions, decoding a "chromosome" string to produce "organelle" strings, but considerable progress has been made with the simplified version. The present arrangement lets us determine the stage at which "coding" becomes a major issue.)

■ Echo's agents have less capability than the adaptive agents described in Chapter 2. Individual agents in Echo do have stimulus-response reactions, implemented by conditions, and they do make extensive use of tags. Individual agents do *not* have the message-passing capabilities required for sophisticated internal models such as default hierarchies. Moreover, the tags control interactions in a much more direct and concrete fashion. Because they are not attached to messages, they do not exhibit the subtle, protosymbolic functions of messages. These simplifications should force the agents in Echo to develop information-processing capacities through more primitive mechanisms. I would like to see the agents evolve programming "languages," rather than supply them with a full-fledged language (the classifier system) at the outset.

If all works well, we will see multiagents in Echo develop detectors and effectors—means of encoding the environment—in coordination with the means for processing this information—programming capability. Each capacity should increase to take advantage of opportunities offered by the other. I would expect to see these capacities exploited by increases in the complexity of interior boundaries in multiagents. Multiagent structure, as defined here, is quite explicit and easy to observe. In a full-fledged classifier system the structure is implicit in the clusters of rules triggered by the different tagged messages. For many *cas* investigations, the more sophisticated internal models possible with a classifier system may be critical; however, Echo's agents offer a simpler approach to questions of diversity and the emergence of organization. Experiments with multiagents have not been run, but the next chapter discusses the possibilities and connects them to experiments that have been performed.

· 4 ·

Simulating Echo

AT THIS POINT we have a description of the mechanisms and interactions that are the foundations for the advanced Echo models. I have two objectives in this chapter. I want to present a speculative scenario that suggests how single free agents can evolve into multiagents, and then into specific aggregates of multiagents generated from a single seed multiagent. Afterward I will discuss the procedure for turning model 6 into a coherent simulation.

A Scenario for the Emergence of Organization

The scenario begins with multiple copies of a free agent that reproduces upon collecting sufficient resources (see Figure 4.1). The agent has neither conditions nor the tags they consult. Under the conventions adopted in Echo, lack of conditions implies a "don't care" (accepts all) condition and lack of tags implies a zero match score, so the agents will still interact. It is up to subsequent crossovers and mutations to originate conditions and generate tags. Thus, the question of whether conditions and tags are useful is still open. If tags and tag-based interactions appear and persist, we will have established a role for them in the emergence of organization, at least in the context provided by Echo.

The first step toward greater diversity would be a mutation giving rise to a conditional mating frame. Crossover and recombination then would have an enhanced role, to exploit the increasing range of combinations possible as mutations accumulate. (We can augment this process by taking another page from the book of genetics, introducing *intrachromosomal duplication*. In its simplest form, this process simply takes a portion of the chromosome and duplicates it, producing a new chromosome with some part doubled. The added part provides fodder for subsequent recombinations and mutations that extend the agent's capabilities.)

More complex organizations begin to emerge when crossover and mutation give rise to conditional adhesion tags. When one of these tags

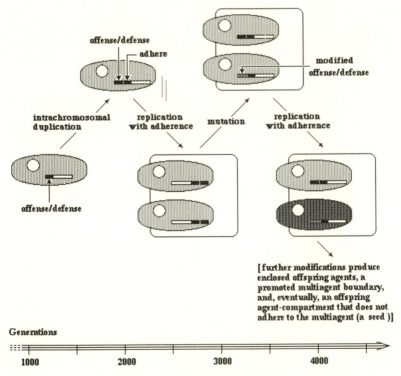

Figure 4.1   Scenario for the Evolution of Multiagents.

is such that the agent's offspring adhere to it, we have the start of a kind of colonial aggregate (like the sponges) that arises from a single agent. Further modifications can produce adhesion scores that force some offspring to form interior boundaries, which cause layering and further organizational complexities. Because agents in the interior face a different environment from agents on the exterior, opportunities for specialization occur. For example, the exterior agents can concentrate on offense, defense, and trade, while the interior agents can specialize in transforming abundant resources into others that are in short supply.

Once such aggregates begin to form, a mutation can move the multiagent boundary marker "upward" in the aggregate, to form a multiagent that includes several agents. The "chromosome" of the multiagent then describes an organization where the included agents serve as component compartments ("organelles"). The shared resources of the agent-compartments provide further opportunities for specialization and reproductive advantage.

At this point recombination and mutation can cause enough differences that, under conditional replication, the offspring of multiagents contain different operational agent-compartments. Thus we obtain an aggregate with differentiated multiagents, even though all the multiagents in this aggregate have the same chromosome. These variations can lead to differences in adhesion. It is even possible for one of the offspring to lose all adhesion to members of the aggregate and be expelled as a free multiagent.

If such an expelled multiagent has the same structure (the same chromosome and active agent-compartments) as the parent that founded the aggregate, the cycle is closed. The evicted free agent becomes a seed that produces adherent offspring that aggregate to yield a new copy of the original aggregate. This process is similar to the one whereby a metazoan is generated by successive divisions of a fertilized egg, ultimately producing a new fertilized egg that can repeat the process.

The appearance of new levels of organization in this evolution depends on one critical ability: each new level must collect and protect resources in a way that outweighs the increased cost of a more complex structure. If the seeded aggregate collects resources rapidly enough to "pay" for the structural complexity, the seed will spread. In Echo, we see new possibilities for further evolutionary modification of the aggregate, through modifications of the seed.

If evolution in Echo were to proceed at all along the lines of this scenario, we would have a rigorous exhibit of the emergence of organization. There is no guarantee that any real system evolves in this way, yet it offers an advantage similar to von Neumann's (1966) rigorous demonstration of a self-reproducing machine. Prior to his work, the possibility of such a machine had been debated for centuries. Von Neumann settled the matter by demonstrating a machine (albeit a simulated machine) that could reproduce itself. Similarly, if some version of this scenario emerges from our simulation, Echo could show that the mechanisms it employs are sufficient to generate sophisticated morphogenesis.

Because the mechanisms at the base of this scenario are few and designed to apply to all complex adaptive systems, we gain a great deal more than just a demonstration of morphogenesis. Tests already completed make diversity an almost certain consequence. That offers an explanation, using common mechanisms, of the pervasiveness of diversity in *cas*. More than that, we gain a uniform description of the processes of learning and adaptation, which brings us much closer to a rigorous framework for describing salient *cas* phenomena.

Just what are the chances of observing this whole scenario, or something like it, in a computer implementation of Echo? Frankly, I do not know. But the scenario is not a naive guess. Many parts of the Echo models have already been tested, and portions of this scenario have been observed. Let us now examine ways of embodying Echo's mechanisms in computer simulations, including those that yield the tests and observations so far completed.

The Nature of Simulation

It will be useful, I think, to start with a bit of stage setting. Most of us are familiar with the use of computers for word processing, spreadsheets, tax calculations, and the like. The less-familiar use of computers for simulation actually goes back to their origin. In a classic paper still worth reading, Turing (1937) shows how to construct a computer, a *universal computer*, that can imitate any other computing machine or computation. The use of computers as devices for imitating other devices is central to the concept of computer-based thought experiments, so it is important to distinguish this use from "number crunching."

The word "simulation" (Latin, "to feign," "to look or act like") itself provides a clue. The heart of a simulation is a map that links parts of the process being simulated to parts of the calculation called *subroutines*. The map has two pieces: (1) a fixed correspondence that relates states of the process to numbers in the calculation, and (2) a set of "laws" that relate the dynamics of the process to the progress of the calculation. A closer look at these two pieces will pay dividends when we come to the specifics of the Echo simulation.

The usual approach to simulation is to divide the process being simulated into components. Then a fixed correspondence is set up, linking the possible states of each part to a range of numbers, as with mathematical models. For example, if we were trying to ascertain the current state of an automobile or airplane, we would ask questions such as, How much fuel is in the tank? What is the rate of fuel use? What is the current velocity? What is the air resistance at this velocity? All of these numbers, and others, would be pertinent to the simulation. When the collection of numbers is sufficient to describe all *relevant* aspects of the process, we say the collection describes the *state of the process*. This piece of the map, then, links the collection of numbers that describes the state of the process to a corresponding collection of numbers in the computer.

The second piece of the map provides the pivotal characteristic of a simulation: it describes how the state of the process changes over time. The computation still uses numbers, but now they relate to a dynamic process. Changes in the numbers reflect changes in the process being simulated. In setting down the laws that determine this part of the map, we take advantage of the computer's ability to execute conditional instructions: IF (the numbers have such-and-such values) THEN (carry out calculation 1) OTHERWISE (carry out calculation 2). We discussed this same ability in the definition of adaptive agents (Chapter 2). As we pointed out, any law that can be specified with logical rigor can be specified by sequencing IF/THEN conditionals with suitable arithmetic operations. This part of the map, then, depends on recasting the mechanisms that generate the dynamics of the process into IF/THEN subroutines in the calculation.

Let me make a general point about simulation and model building: in building the map that defines a model, selection is critical. The cartooning metaphor is once again instructive. The model (cartoon) can be more, or less, faithful to the original and, as always, which it is depends on the purpose of the model (cartoon). We may opt for simplicity, or even distorted similarity, at a cost in faithfulness in order to emphasize some basic element. Newton, in building his models, ignored friction in order to get a more definitive look at momentum. His slightly unfaithful model emphasizes the principle that "bodies in motion persist in that motion, unless perturbed by forces." Aristotle's earlier, more faithful model implicitly included friction, leading him to enunciate the "basic principle" that "all bodies come to rest." Aristotle's model, though closer to everyday observations, clouded studies of the natural world for almost two millennia. Model building is the art of selecting those aspects of a process that are relevant to the question being asked. As with any art, this selection is guided by taste, elegance, and metaphor; it is a matter of induction, rather than deduction. High science depends on this art.

An Echo Simulation

The heart of the Echo models is the interaction between agents at a site, so that becomes the central routine in the simulation (see Figure 4.2). The design is such that we can carry out computations as if the populations at each site undergo their interactions simultaneously, much like a group of billiard balls simultaneously in motion on a billiard table.

On the billiard table, at any given time, some balls will be colliding while others will be in free trajectory. Similarly in Echo, some agents will be interacting at any given time and others will not be involved. The first step in the central routine, then, is to determine the interactions at a site. The simplest approach is to assume that, at that site, agents come into contact at random. That notion can be implemented simply by picking pairs of agents at random from the list of agents at the site.

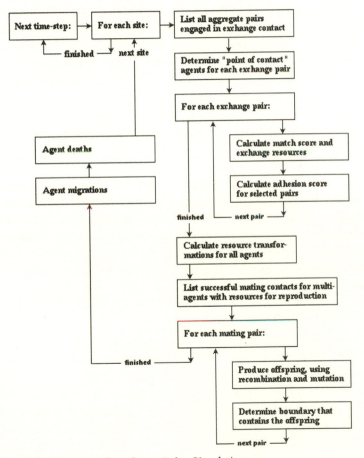

Figure 4.2   Program Flow for an Echo Simulation.

Note that a contact does *not* mean that an interaction will necessarily take place—that depends on the conditions and match scores involved. Contacts only set the stage for interaction.

This notion of contact has to be extended to allow for interactions between aggregates. The general principle, enunciated earlier, is that all interactions are ultimately between the individual agents in the aggregate. The easiest way to make this extension is to select one agent at random from the list of all agents at the site, then select the second agent at random from within the domain of interaction of the first agent.

It is useful conceptually to divide all contacts into two types. One type, which I'll call *exchange contact*, involves exchange interactions and adhesion interactions that are *not* between parent and offspring. The pairs used in exchange contacts are drawn at random from the general population, subject only to conditions set by agent boundaries. The second type of contact, which I'll call *mating contact*, involves mating and the adhesion of offspring. The list of candidates in this second case is restricted to those members of the population that have collected enough resources to reproduce. That is, the list of mating candidates consists of multiagents with enough resources to reproduce the whole of the multiagent's chromosome (recalling that a free primitive agent is, formally, a single-agent, single-boundary multiagent). As with exchange contacts, the pairs are drawn with due attention to the domains of interaction imposed by boundaries.

The simulation checks all exchange contacts, then it checks all mating contacts. We'll look at each in turn.

EXCHANGE CONTACTS

For contacts of the first type, exchange conditions are checked first. By the procedures detailed in model 2 (*Conditional Exchange*), the exchange condition of each agent is checked against the offense tag of the other. If the exchange conditions of both agents are satisfied, then each offense tag is matched against the defense tag of the other agent and match scores are calculated. Resources are exchanged according to the specifications of model 1 (*Offense, Defense, and a Reservoir*). If one condition, but not the other, is satisfied, then the agent with the unsatisfied condition has a chance of aborting the interaction; otherwise the interaction proceeds as when both conditions are satisfied. If neither exchange condition is satisfied, the exchange interaction is aborted.

Once the exchange interactions are completed, some of the pairs, chosen at random from the set of executed exchanges, undergo a test for adhesion. The proportion of pairs so chosen is open to the experimenter; it is a parameter of the model. (Adhesion under these circumstances allows formation of aggregates from members-at-large in the population, in contrast to adhesion that occurs between offspring and parents under mating contacts.) For each chosen pair, the adhesion tag of each agent is matched against the offense tag of the other agent in the pair. Net match scores are calculated and boundaries are adjusted according to the result, as detailed under model 4 (*Adhesion*).

When an exchange contact results in resource exchange or adhesion between agent-compartments, then those agent-compartments are marked *active* for later use with conditional replication (see below).

MATING CONTACTS

Mating contacts are restricted to multiagents that have accumulated enough resources in the reservoirs of their compartment-agents to allow replication of all their compartment-agents.

Because mating contacts are centered on multiagents, we need to determine which mating condition to use when the multiagent has more than one agent-compartment. Intuitively, it would seem natural to restrict mate detection to agents in the outer boundary of the multiagent. The simplest resolution apparently is to select one of these agents at random, at each contact, as a determiner of the mating condition. That is, each time there is a mating contact between two multiagents, one of the agents in each of the outer boundaries is used to determine whether or not a mating interaction follows the contact. Note that the multiagent can present different "faces" on successive contacts if there are several agents in the outer boundary with different mating conditions.

Once the determining agents have been selected in each multiagent, the procedure detailed in model 5 (*Selective Mating*) is used to determine whether or not a mating interaction ensues. The mating condition of each agent is checked against the interaction complement substring in the other agent's tag segment. The contact turns into a mating interaction only if the mating conditions of both agents are satisfied.

A mating interaction proceeds in the usual manner for genetic algorithms, with a pair of offspring being produced from the parents. The chromosomes of the two parent multiagents are copied, crossed, and mutated, producing two offspring. (This is only vaguely similar to the real biological process, but it does exploit recombination of discovered building blocks, a vital feature of *cas*. It is easy to bring the process closer to biological reality, but at the cost of additional complexity in computation.)

Once the offspring are produced, each is "assigned" to one of the parents to test for mutual adhesion. Then the adhesion tags, specified in the control segments of parent and offspring, are matched and scored, as detailed in model 4 (*Adhesion*). This step makes possible a kind of morphogenesis, producing aggregates through adhesion of successive generations of offspring. As successive generations are produced, the complexity of the aggregate can increase through two mechanisms:

(1) The calculated match score can force the offspring, or the parent, to move interior to the boundary containing the parent; if no interior boundary exists, it can force the formation of a new boundary.

(2) The conditional replication conditions, discussed in model 6 (*Conditional Replication*), may dictate that certain agent-compartments in the offspring multiagent be "off" (effectively absent). It is at this point that the active/inactive status of compartments, set during the exchange contacts, comes into play. The conditional replication condition of each agent-compartment in the multiagent is checked against the interaction tags of the multiagent's active agent-compartments. An agent-compartment is "on" (present) in the offspring only if the replication condition is satisfied, as detailed under model 6. Because only agents that are on can interact, conditional replication can substantially alter the patterns of exchange and adhesion as successive generations are produced. This is the stage at which some multiagent offspring can be

set free from the aggregate, through lack of adhesion, and we have the possibility of producing a seed that will generate a whole new copy of the aggregate.

A Flow Diagram

The foregoing interactions between agents constitute the heart of Echo, but there are still some "housekeeping" activities. They include absorption of resources from the site, resource transformation, agent death, and migration from site to site. I'll fit each of them into the flow diagram for the Echo simulation (Figure 4.2).

Absorption of resources from the site is most easily handled if we consider the site itself to be an agent with a tag. Then a conventional agent residing at that site can interact with the site, if it has an appropriate offense tag and exchange condition. Under this arrangement, the ability of an agent to absorb resources from the site can evolve through changes in its tag and condition, and the whole process simply becomes part of the exchange contact section of the simulation.

Resource transformation is contingent on the presence of an appropriate section in the agent's chromosome (details under model 3). It can be executed at the end of the exchange contacts as a precursor to the mating contacts.

Agent migration is most easily executed at the end of all contacts. In the simulation, each agent is assigned a site label (coordinate), and migration consists of changing that label to the label of an adjacent site. In the simplest case a few agents are selected at random to have their site labels changed. A more realistic version would have the probability of selection for migration increase if the agent's reservoir were low in critical resources. (There are many variations on this theme.)

Agent death (as outlined in model 5, *Selective Mating*) can be the last activity of each time-step. In the simplest case each agent has a fixed probability of deletion. This process can be made more realistic by charging each agent a "maintenance cost" on each time-step, say one unit of each resource that it uses in its chromosome. If the agent's reservoir is devoid of *all* such resources after the charge, then it has an increased probability of deletion. (Again, there are many variations on this theme. Note that when maintenance costs are charged, there is an advantage if the parent passes some of the resources in its reservoir on to its offspring.)

Tests: A Population-Based Prisoner's Dilemma

At the time of this writing, only Echo model 1 has undergone extensive tests. There is sophisticated software with good provision for interaction and flexible means of displaying the action, much as one would expect of a flight simulator (to be discussed later in this chapter). We have observed biological arms races (see Figure 1.12), and situations such as the caterpillar-ant-fly triangle have been tested.

Extensive tests of the other models lie in the future. We do have results, from an Echo-like simulation, on the effects of tags in breaking symmetries; these are interesting enough to warrant discussion.

At the end of Chapter 2, I introduced the Prisoner's Dilemma to illustrate the ways in which an adaptive agent improves its strategy. That example can be extended easily to a population of agents in an Echo-like environment. As in a billiard ball model, agents come into contact at random and each has a strategy that it acquires from its parents through recombination and mutation. When two agents come into contact, they execute one play of the Prisoner's Dilemma, each acting as dictated by its strategy (see Figure 4.3). Over successive plays each agent accumulates the payoffs it receives, and it produces offspring at a rate proportional to its rate of accumulation. (This is a simplified version of the Echo format in that the agents have an explicit fitness function, with no need to collect resources to "spell out" their strategies.) The object is to observe what strategies the agents evolve over time as they adapt to each other.

Within this format let's look at two experiments. In one experiment, each agent has a chromosome that specifies its strategy, but it has no means of distinguishing agents from one another. It is as if the agents were all cue balls on a billiard table, with hidden internal models (strategies). In the other experiment, each agent has a chromosome that specifies an exterior tag and a condition for interaction, as well as a strategy. There is no necessary connection between the tag, the condition, and the strategy. All are separate parts of the chromosome, and all are open to separate adaptations. The experiments, then, present two worlds, one with tags and one without.

Will there be consistent differences in the strategies that evolve in these two worlds? From our earlier discussions, we would expect an advantage from the symmetry breaking provided by the tags. For example, an agent developing a condition that identifies tags associated with "cooperators" will prosper from the increased payoff that results. We'll see that experiment does indeed bear out this conjecture, even as it reveals some additional twists.

Some earlier experiments on selective mating (Perry, 1984) bear on this process. Consider a population with a variety of randomly assigned tags, and selective mating conditions that examine those tags. The number of ways of combining tags with conditions grows rapidly as the

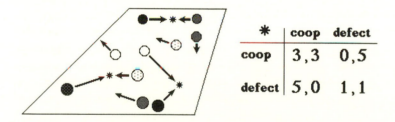

✻	**coop**	**defect**
coop	3,3	0,5
defect	5,0	1,1

- **Each agent has a strategy determined by a set of rules.**

    For example, one of ●'s rules could be: IF {○} THEN coop.

- **At each random contact (✻), the pair of agents involved plays one round of the Prisoner's Dilemma.**

- **Agents accumulate the payoff that results from successive plays of the game.**

- **When an agent's accumulated payoff exceeds a predetermined threshold, it reproduces itself (with mutation).**

Figure 4.3   A Population-Based Version of the Prisoner's Dilemma.

number of tags and conditions increases. Even with modest numbers, it is likely that some tag/condition combination will confer a slight reproductive advantage. For example, a combination can restrict mating to "compatible" individuals that have building blocks that work well together, thereby producing fewer ill-adapted offspring under crossover. Any early, accidental association of a tag with a trait that confers a reproductive advantage will spread rapidly because of the higher reproduction rate. Tags that are originally meaningless, because of the random assignment, then acquire meaning. They come to stand for particular kinds of compatibility. Evolutionary processes refine selective mating conditions based on these tags, so that agents can react to this compatibility and thereby increase their fitness. In Perry's experiments different sites offered different possibilities for building blocks and compatibilities. The amplification of tags and tuning of conditions, under a genetic algorithm, led to well-defined, site-specific species that did not crossbreed.

We would expect similar advantages to accrue to agents using tags in the population-based Prisoner's Dilemma experiment: an agent developing a condition that identifies tags associated with cooperators should prosper from the increased payoff that results. As in the selective mating experiments, there is strong selection for combinations of tags and conditions that favor profitable interactions. In effect, the agents develop tacit models, anticipating the effects of interacting with agents that have certain kinds of tags.

Rick Riolo, at the University of Michigan, has executed experiments along the lines just described. They confirm the expectation that tags provide an advantage, and they yield some interesting insights.

Consider first the agents without tags. At each contact between a pair of agents, one play of the Prisoner's Dilemma is executed. Because the pairing is random, the opponents are random and unidentified. There is no basis for implementing conditional interactions. In this evolving population, the productive tit-for-tat strategy never establishes itself for any extended period. Interactions are largely of the minimax defect-defect kind, clearly disadvantageous relative to cooperate-cooperate interactions.

Agents *with* tags evolve along an entirely different path. At some point, as the strategies evolve, an agent appears that (1) employs tit for tat, and (2) has a conditional interaction rule based on a tag carried by a subpopulation that is susceptible to tit for tat. That is, the agent restricts its interactions to agents having strategies that (often) produce a cooperate-cooperate result under tit for tat. The resulting higher reproduction rate causes both this agent and its cooperating partners to spread through the population. Subsequent recombinations provide tit-for-tat agents that restrict their interactions to other agents playing tit for tat. Once established, such a subpopulation is highly resistant to invasion by other strategies. In biological terms, these agents, with their conditional tag-mediated interactions, have found something close to an evolutionarily stable strategy. (The notion of an evolutionarily stable strategy, ESS, was introduced by Maynard Smith, 1978. Such a strategy, once established in a population, resists the invasion of all other strategies that might be introduced, in small numbers, by evolution.)

Even in the limited confines of the population-based Prisoner's Dilemma, the evolutionary opportunities for adaptive agents with tags go considerably beyond the ESS just mentioned. For example, mimicry becomes possible. While pursuing a different strategy, an agent can present the tag associated with tit for tat. The presence of an agent with a tag that has a well-defined functional meaning—tit for tat in this case—opens new niches for other agents. These niches are usually constrained in size, depending as they do on the continued presence of the "founding" agent. In mimicry, biological studies suggest, the mimic can only occupy a small proportion of the overall population relative to the agents being mimicked. This is so because the other agents begin to adjust to the deception as the proportion of mimics increases. Negative feedback sets a limit on the mimic's expansion. It is typical that tags provide niches of limited "carrying capacity," leading to highly diverse systems with no "superindividual" that can outcompete all comers.

Future Uses

There are two broad lines of development open to Echo. One involves increasingly sophisticated thought experiments aimed at an understanding of the mechanisms and principles of *cas* evolution. The other adds more realistic elements to Echo so that it can be used as a "flight simulator" for policies directed at complex adaptive systems.

THOUGHT EXPERIMENTS

The scenario for the emergence of organization (at the beginning of this chapter) is an example of what we can expect from thought experiments based on Echo. Results already obtained with model 1, and partial tests of some of the other mechanisms such as conditional exchange, give credence to that scenario. But the outcome is far from assured, and there is much to learn along the way.

It is worth emphasizing again that such computer-based thought experiments are not attempts to match data. They are, rather, an attempt to discover the adequacy of particular mechanisms. It is not easy to come up with *any* explanations for *cas* phenomena, let alone candidates that can be reduced to rigorous models. (As C. S. Pierce would say, they are not so plentiful as blueberries; see Wiener, 1958.) So it is an advance just to locate possibilities. It is useful to know how far we can go with specific mechanisms, and the details of our failures may suggest new mechanisms.

When mechanisms do prove adequate to generate parts of the scenario, *then* it is worthwhile to see if they exist and play similar roles in real *cas*. Successful thought experiments suggest where to look in the complex tangle of possibilities and data, and they provide guidelines for new experiments. When we reach this stage, the thought experiment approach begins to merge with the flight simulator approach.

FLIGHT SIMULATORS

The copilot of a large commercial aircraft may have less than an hour of actual flight time on that particular class of plane (say a 747) at the time of his or her first flight with passengers. What the copilot *does* have is many, many hours on a flight simulator for that class. It might seem that the balance of time should be the other way around, but I prefer it the way it is. In a simulator, a pilot can experiment in a way that would be infeasible with real aircraft, let alone an aircraft with passengers. The pilot can test performance with a two-engine flameout, or recovery from inverted flight. There have been cases where such experience has saved lives, as a few years ago, when all the control surfaces on a passenger plane became disconnected. The plane was landed by a pilot who had tested, on a simulator, the ability to maneuver a plane on thrust changes alone.

Of course, the value of simulator experience hinges on the simulator's faithfulness to the aircraft it models. To be useful, the flight simulator must successfully mimic the real plane under the full range of events that can occur. Solid theories of aerodynamics and control, a natural cockpit-like interface, and superb programming are vital ingredients of an acceptable flight simulator. Given this complex mix, how is one to validate the resulting simulator? Even relatively simple programs have subtle bugs, and flight simulator programs are far from simple.

Enter the experienced pilot. The pilot "takes the simulator out" for a series of test flights, performing the maneuvers suggested by long experience with real aircraft. In particular, the pilot "pushes the envelope," taking the simulated plane close to the edges of its design parameters. If the simulator performs as the pilot expects, we have a reality check; if not, back to the drawing board. It's possible that there is some unusual, untested pocket where the simulator departs from real performance (similar surprises are possible with real aircraft), but it is unlikely that the simulator is systematically wrong if it passes such a "wringing out."

This means of attaining a reality check sets a goal for simulations that mimic real systems. Individuals experienced with true *cas* should be able to observe familiar results when executing familiar actions in the simulator. This puts a requirement not only on the programming, but also on the interface provided. We should not expect the tester to become an expert in the simulation program, any more than we expect the pilot to be an expert in the programming behind the flight simulator. The pilot was provided with a cockpit and display that enabled him to take familiar actions and observe the results in a familiar way. An expert ecologist, or economist, or politician should have similar advantages when dealing with a simulation like Echo, when it is to be used to mimic reality.

Providing a realistic interface is a difficult and unusual task when we're dealing with *cas*, but the interfaces of some of the more interesting "political" video games point the way. For example, SimCity (Wright, 1989) provides intuitive, natural ways of looking at, and responding to, an urban situation that involves taxation, zoning, crime, votes for office, and so on. The game itself greatly simplifies urban dynamics, but the interface is substantially more sophisticated than that provided for extant simulations in the *cas* arena.

The end point, a *cas* simulation with a realistic interface, is highly desirable, because it enables an ecologist, or economist, or politician to try out alternatives that could not possibly be tried in real systems. Intuition can be augmented by detailed exploration of the effects of alternative courses of action. As for the pilot, ways of controlling disaster scenarios can be tested. With sufficient forethought, disasters can even be used in a positive way to change habits. In the aftermath of the 1994 San Francisco earthquake, as much as 80 percent of the local population started using the public transportation system. After a few months ridership slacked off to something close to previous levels, but it need not have. The increase in ridership was a predictable consequence of the disaster, and a tremendous opportunity. Some prior thought about reinforcing the change would certainly have made it possible to retain a large proportion of the increased ridership.

How Far Have We Come?

We now have a way of modeling adaptive agents, and we have a way of investigating their interactions. The models proposed are by no means the only ones that could have been set up. Different ways of looking at *cas* inevitably lead to different emphases and different models. For all of that, the models here are not arbitrary.

The most important constraint is a requirement that the computer-based model be something more than a programming language that can define all agent strategies. Just because a language has the power to describe a phenomenon does *not* mean that it will provide useful insights. Two languages that have the same formal capabilities may provide very different insights. The model, and the language it uses, must be tuned to the phenomena and questions of interest.

To better understand this, we need to take a closer look at what we mean when we say that two sets of assumptions, say two axiom systems for geometry, are *formally equivalent*. They are formally equivalent when all the logical consequences, the *theorems*, of one system are identical to those of the other. It is often possible to establish the formal equivalence of two systems without knowing much about the theorems they entail. This can be of considerable advantage in showing us that our formalization has not undershot the mark by being insufficiently powerful. Yet it is not enough for present purposes. Different formally equivalent systems can pose substantially different difficulties when it comes to deriving key theorems. They may have quite different "accessible" expressiveness.

Consider two formally equivalent formulations (axiom systems) for Euclidian geometry. In one, the shortest proof of some important theorem, say the Pythagorean, requires less than twenty steps, while in the other the same theorem requires at least a billion steps (or any number you care to choose). We know that such differences exist in formally equivalent systems because of theoretical work done in the first third of the twentieth century (see Mostowski, 1952). Certainly these two systems will offer different insights into Euclidian geometry, for any feasible amount of effort. That is, formally equivalent does not mean "equivalent with respect to accessible insights." If we have selected questions in mind, it is not enough to establish that a formalism is formally adequate for answering those questions. A close look at the questions is indispensable for arriving at a rigorous presentation that will aid, rather than hinder, the investigation.

Applied to adaptive agents, these strictures validate the point made at the start of this section. We require more than a programming language that has the formal power to express all adaptive agent interactions. Adaptive agents come in startling variety, and their strategies are correspondingly diverse, so we need a language powerful enough to define the feasible strategies for these agents. But that is just a beginning. Models that can advance our understanding of questions about diversity, internal models, lever points, and the like, must satisfy additional strong constraints. We must look at the activities of the adaptive agents—performance, credit assignment, and rule discovery—and tailor the model for a direct investigation of the interactions that arise from these activities. And we must provide well-defined evolutionary procedures that enable agents to acquire learned anticipations and innovations. These constraints are so powerful that it is not easy to come up with *any* rigorous model that exhibits these capacities, let alone one that is plausible.

Echo does satisfy these constraints and it is, to a degree, plausible. Simulator runs with the simpler Echo models have exhibited the kinds of evolution and interaction that we observe in real *cas*. Preliminary runs that utilize some of the more sophisticated mechanisms have also shown the enhancements we would expect from those mechanisms. And several projects, some simple, some complex, are modifying Echo to use real data. But there is a long way to go.

On a broader scale, I have no doubt that thought experiments, guided by simulations such as Echo, are vital to a general understanding of complex adaptive systems. We need the halfway house provided by such simulations. The traditional direct bridge between theory and controlled experiment is all but impossible in this situation. We cannot follow the traditional experimental path, varying selected variables under repeated runs, while holding most variables fixed, because controlled restarts are not possible with most *cas*, and because some *cas* operate over long time spans. The computer-based models can give us this possibility if they capture the "right" aspects of real *cas*. In this the models are no different from the designed experiments: Selection guided by taste and experience is crucial. In the end, simulations such as Echo will be productive only if they suggest patterns and building blocks that can be turned into the stuff of mathematical theory.

· 5 ·

Toward Theory

ALMOST ALL OF OUR EFFORT to this point has been spent in getting to, and designing, the halfway house represented by Echo. Now we look to the destination—general principles. Although that destination is still on the horizon, there are useful landmarks, and those of us who have been studying *cas* at the Santa Fe Institute are optimistic about the way ahead. We believe that there *are* general principles that will deepen our understanding of *all* complex adaptive systems. At present we can only see fragments of those principles, and the focus shifts from time to time; but we can see outlines, and we can make useful conjectures. Just what can we see and imagine?

Mathematics is our sine qua non on this part of the journey. Fortunately, we need not delve into the details to describe the form of the mathematics and what it can contribute; the details will probably change anyhow, as we close in on the destination. Mathematics has a critical role because it alone enables us to formulate *rigorous* generalizations, or principles. Neither physical experiments nor computer-based experiments, on their own, can provide such generalizations. Physical experiments usually are limited to supplying input and constraints for rigorous models, because the experiments themselves are rarely described in a language that permits deductive exploration. Computer-based experiments have rigorous descriptions, but they deal only in specifics. A well-designed mathematical model, on the other hand, generalizes the particulars revealed by physical experiments, computer-based models, and interdisciplinary comparisons. Furthermore, the tools of mathematics provide rigorous derivations and predictions applicable to all *cas*. Only mathematics can take us the full distance.

The Separation between Observation and Theory

To see more clearly the distance between observation and theory for *cas*, let's look again at an example—this time concerning sustainability.

Early in this century the supposedly inexhaustible forests of the Upper Peninsula of Michigan were cut down, reducing most of the area to a barren stumpland. Then, during the depression of the 1930's, the Civilian Conservation Corps (CCC) was formed to reduce the devastating effects of unemployment in the cities. Over several years, at a surprisingly low cost to the government, the CCC (many of whose members in this region were drawn from Detroit) planted seedlings throughout vast tracts of the Upper Peninsula. Now, half a century later, the land is once again forested, to the great benefit of tourism and the lumber industry (more cautious this time around). Extensive interviews of former CCC members several decades later show that almost all of them look on this period as a turning point in their lives.

We would seem to have here a prime example of a lever point in a political-economic context. But questions abound. Would this procedure be repeatable, at least in outline, if we replaced Detroit and the Upper Peninsula with Los Angeles and the forests of the Northwest? Is this an example of a broader class of symbiotic solutions coupling inner-city problems with resource sustainability? More generally, what combined circumstances in economics and politics make such long-horizon investments possible? Must they always be centered on some disaster, as in our earlier example of the San Francisco earthquake and public transport? Why do those working with renewable resources, such as forests and fish, exhaust those resources when they know (as they do) that the action destroys their livelihood? Is this somehow connected with the downside of the Prisoner's Dilemma?

The last two of these questions have anecdotal answers. We talk of the "tragedy of the commons," where some common resource is overrapidly exploited by everyone, because each person mistrusts the moderation of others. That is indeed reminiscent of the defect-defect solution of the Prisoner's Dilemma. And we talk of the "mobility of capital," where the investors in an industry are distinct from the "locals" (the workers and owners), so the investors simply reinvest in some other industry when the local industry collapses. The investors don't suffer the consequences of the collapse, at least in the short run, so they show little concern. These answers have more substance than, say, the pundits' reasons for the rise or fall of today's stock market, but we have no firm basis for knowing when, or if, they apply.

We could, with substantial effort, model situations like this in Echo. A flight simulator version would be particularly helpful, letting the politician or economist observe the short-term and long-term outcomes of policies they consider feasible. Still, that is not really enough. We would do much better with guidelines that suggest where to look. We need some way of searching beyond familiar policies, which may offer little or may be caught in a legislative deadlock. The space of possible policies is large, and there may be some that exploit lever points, if we can just uncover them. But lever points, at least in our examples, are often obscure and not easily located by trial-and-error exploration. In these cases, theoretical guidelines relating lever points to specifics of the problem would be an invaluable help.

Two-Tiered Models

The first step in moving toward an appropriate theory is, once more, careful selection of mechanisms and properties from a multitude of possibilities. It is helpful to recast the problem in a framework, such as Echo, that relies on selected mechanisms common to all *cas*. It is particularly helpful if the model is kept simple, while retaining salient features of the problem that aim at thought experiments rather than a full flight simulator. We can still keep looking toward theory, favoring elements that can be mathematicized, where this can be done without jeopardizing relevance.

Consider the CCC example. A major part of the simulation in Echo would center on the action of one set of agents (inner-city workers) as catalysts for the recovery of another set of agents (the trees), after the first set had moved from one site (Detroit) to another (the Upper Peninsula). Here we are dealing with the consequences of *flows* (Chapter 1). We are also dealing with differing timescales. The workers move and act on one timescale, call it a "fast dynamic," while the trees recover on a much longer timescale, a "slow dynamic."

With the help of Echo, we can recast the problem in terms of flows of resources between different kinds of agents, as is true of most *cas* problems. We can make solid contact with mathematical models if we make two simplifying assumptions: (1) the agents can be usefully aggregated into species or kinds, and (2) there is a rapid mixing of resources among agents of like kind. With respect to the first assumption, the hierarchical organization typical of *cas* usually makes aggregation easy and natural. (See, for example, the discussion of default hierarchies in Chapter 2.) The second assumption assures that the consequences of interactions are rapidly distributed within each aggregate. Rapid distribution, in turn, assures that we can assign average resource levels to aggregates at each instant, without being stymied by nonlinear effects *within* the aggregate. Under these assumptions we can treat Echo-based models (and complex adaptive systems) in a kind of two-tiered format.

The Lower Tier

The lower tier concerns itself with the flow of resources between agents of different kinds. The combination of rapid mixing within each kind, and random contact between kinds, makes possible a mathematical model much like the billiard ball model discussed in the first chapter. That is, we can treat each *kind* of agent as a *kind* of billiard ball, and for each pair we can determine a reaction rate. The rate is directly determined by the exchange condition and the exchange scoring mechanism specified for each agent in Echo (see model 2 in Chapter 3). The result is an array of reaction rates (see *Nonlinearity* in Chapter 1).

Once this array has been computed, we are close to having a mathematical model that describes changes in flow over time. In particular, we are close to describing mathematically the change in the proportion of each kind of agent at a site, as time elapses. The relevant vehicle is the version of the Lotka-Volterra equations discussed in the nonlinearity example. Those equations let us determine the changes in *proportion* of each agent-kind by using the reaction rates for various possible pairs. However, we face a problem. The flow model gives the *total* resources held by each agent-kind, but the equations require the *proportion* of each agent-kind. Different kinds of agents use different amounts of resources in their structures, so aggregate resource totals do not directly determine agent-kind proportions. To derive the proportions, we must divide the aggregate resource totals by the amounts of each resource required to make a copy of that kind of agent.

The rapid mixing assumption now lets us treat the resource totals as equally shared by the individuals in each aggregate. Specifically, the rapid mixing assumption ensures that all reservoirs in the aggregate hold similar amounts of each resource. Knowing this, we can determine the number of agents in the aggregate by dividing the total resources held by the number of each kind of resource required to build that agent's chromosome. Then, knowing the *number* of individuals of each kind, we can determine their *proportions* in the total of all individuals. Having determined the proportion, we can use the Lotka-Volterra equations as a mathematical description of the changing resource flows mediated by the agents.

Even at this preliminary level, some theoretical progress can be made concerning lever points. Because agents can have surpluses of some resources, only certain resources held by the aggregate "count" toward the number of any given agent-kind. The notion of a "bottleneck resource" emerges. A close look at the flow model shows that a change in the bottleneck resource—say a new interaction greatly increases its level—can have much the effect of a mutation. It can open a cascade of new interactions. Changes in a bottleneck resource often give rise to effects far out of proportion to the change.

To adopt a term from physics, the lower tier gives us a mathematical model of the fast dynamics of the system.

THE UPPER TIER

For a mathematical theory of *cas* to be effective, the fast dynamics of the flows must be successfully coupled to the slow dynamics of long-term adaptation and evolution. In this two-tiered model, it is the upper tier that specifies the evolution of the agents. It uses a genetic algorithm to change the structures of offspring, as described at the end of Chapter 2. In Echo the resulting agent structures precisely determine the amounts of resource exchanged, so the reaction rates of the lower tier are directly coupled to the results of actions in the upper tier. Note that a change in the definition of the agent-kinds (aggregations) used in the lower tier will result in different couplings to the upper tier.

In selecting the aggregations and couplings to the lower tier, we want to make it easy to see how the network changes when the genetic algorithm causes given building blocks (schemata) to spread and recombine. One extreme would be to allow one node in the network for every distinct agent. Then the lower tier would be formally correct, but the patterns of change would be spread over large numbers of nodes. At best, the patterns would be difficult to discern. The lower tier only becomes useful, both computationally and theoretically, when we can aggregate agents into kinds based on the presence or absence of the chosen building blocks. Then the patterns of change relative to these building blocks will be manifest. This is the burden of the earlier "useful aggregation" assumption (look back again at Chapter 1).

Aggregation of agents, however, raises a problem similar to our earlier difficulty with aggregation of resources. For a given pair of agents, we can directly determine a flow of resources and a reaction rate (as detailed in Chapter 3). However, this is not necessarily an appropriate reaction rate for the pair of *aggregates* to which these agents belong. Agents of a given kind will not generally exchange resources in identical fashion; after all, we only collected them into a common kind because they had *some* building blocks in common. So two agents of the same kind may have different associated reaction rates. This puts us squarely into the difficulty discussed under the topic of nonlinearity in Chapter 1. We cannot simply average the reaction rates of individuals of a given kind to get a reaction rate for the aggregate agent-kind. That is, reaction rates associated with the flow network are not simply related to reaction rates associated with agent pairs.

We *can* determine a useful reaction rate for an agent-kind if the constituent agents are not too different from one another relative to their ability to exchange the resources of interest. In this instance the individual reaction rates are close to one another, so that the flow calculated with the average rate will not differ greatly from the actual flow. (The actual flow is determined by summing the individual flows of individual agents.) At worst, we can establish that no agent has a reaction rate slower (larger) than a determined amount, allowing us to determine bounds on the flow, rates of reproduction, and the like.

Keeping the individual reaction rates in an aggregate close to one another actually is largely under the control of the theorist setting up the two-tiered model. That person selects the characteristics that group the agents into aggregates. By selecting appropriate characteristics, the theorist can limit the variation in the individual reaction rates within each aggregate. The building blocks of the exchange conditions and the interaction tags are central to this purpose. By aggregating agents with the same alleles for these building blocks, the theorist can assure closeness of reaction rates, while benefiting from a simplified lower tier.

In sum, one way to generate a useful coupling of the upper tier to the lower tier is to aggregate agents with similar building blocks in the parts of the chromosome devoted to the offense tag, the defense tag, and the exchange condition. If we further constrain these aggregates by conditional replication, we achieve something much like biological speciation. Patterns should be sharpened because aggregates cannot blend into one another. In any case, the upper tier has the effect of continually changing the flow network of the lower tier, as the agents evolve and adapt under the genetic algorithm.

A THEORY OF TWO TIERS

The relevant theory for the upper tier starts with the schema theorem for genetic algorithms because that theorem tells us about the spread and decline of building blocks. However, the version of the theorem given at the end of Chapter 3 is only a beginning. We need a version of the schema theorem that holds for the implicit fitness of the Echo models. And the theorem should tell us about the spread of schemata across kinds, with particular attention to the effects of selective mating. This element is important if we are to understand the spread of building blocks in real *cas*, such as the spread of the Krebs energy transformation cycle throughout the vast range of aerobic organisms or the spread of computer chips throughout machines ranging from automotive engines to cameras.

Given the perpetual novelty of agents in the Echo models, we need still more from a satisfactory theory. The unfolding development of an Echo world is a trajectory through a space of multiple possibilities; we need to know something of the form of this trajectory, particularly because *cas* rarely reach end points or equilibria. We are likely to understand a *cas* process only if we know what the trajectory looks like along the way.

It will be difficult, perhaps impossible, to predict details of the trajectory, but surely it is far from a random walk. At worst, we may face a phenomenon similar to the day-to-day, month-to-month changes in weather, though I think *cas* are more predictable than that. Even with the weather, there are building blocks—fronts, highs and lows, jet streams, and so on—and our overall understanding of changes in weather has been much advanced by theory based on those building blocks. It is still difficult to predict detailed weather changes, particularly over an extended period. Nonetheless, theory provides guidelines that lead us through the complexity of atmospheric phenomena. We understand the larger patterns and (many of) their causes, though the detailed trajectory through the space of weather possibilities is perpetually novel. As a result, we can do far better than the old standby: predict that "tomorrow's weather will be like today's" and you stand a 60 percent probability of being correct. A relevant theory for *cas* should do at least as well.

Complex adaptive systems exhibit more regularities than weather for at least two reasons. First, there is the persistence of favored building blocks. (In biological systems, the Krebs cycle is pervasive in both space and time; in economies, taxes too are pervasive in space and time.) Second, there is the phenomenon known in biology as *convergence*, which imposes further predictable regularities. Convergence in this sense should not be confused with the attainment of end points (fixed points), the subject of mathematical convergence. Here convergence refers to the similarity of agents occupying similar niches. With some knowledge of the niche, we can say something of the form of the agent that will occupy it. As an example, biologists recently discovered a tropical flower with a throat of unprecedented depth, a flower belonging to a genus invariably pollinated by moths. The niche provided by this flower led the scientists confidently to predict the existence of a moth, yet to be found, with a proboscis of equally unprecedented length.

The regularities provided by building blocks and (biological) convergence imply regularities in the development of the flow network. These, in turn, imply that agents attain high concentrations at certain kinds of nodes. New variants are most likely to arise where there are many agents; more samples mean more possibilities for variation. Accordingly, the generation of new agent-kinds (nodes) should center on these well-populated nodes, a kind of *adaptive radiation*. So we have some hints about how the network would grow. If the fast dynamic is modeled by a set of equations of the Lotka-Volterra form, this growth means adding new equations to the set. The added equations produce corresponding changes in the dynamics. To couple this growth to the upper tier, we need a version of the schema theorem that takes selective mating into account, while using only endogenous fitness. Such a theorem would let us determine something of the form of the trajectory through the space of lower-tier flow networks. It could give us some idea of what convergence means in this general setting, a setting that holds for all complex adaptive systems.

A Broader View

This two-tiered model undoubtedly captures a substantial portion of what is going on in *cas*. Yet we are only starting to give it the precision required for mathematical theory. Two advances in mathematics would help provide a theory of this two-tiered model. One is an organized theory of a dynamics based on *sets* of equations that change in number (cardinality) over time. Another is a theory that relates generators (building blocks) to hierarchical structure (for example, default hierarchies), strategies (*classes* of moves in games), and the "values" associated with those strategies (game payoff).

Now an aside, for those conversant with mathematics. Such a mathematics would resemble the use of generating functions to estimate parameters of stochastic processes (see Feller, 1950). Its combinatorial aspect would have the flavor of the work on "automatic" (automaton) groups (see Baumslag, 1994). The stochastic aspect can be studied with the help of Markov processes, but the usual treatment of such processes, which concentrates on eigenvectors and fixed points, will *not* be of much help. Instead, we need to know what happens to aggregates during the transient part of the process. Aggregation of states of the full process encounters the usual difficulties with nonlinearities; still, there are ways around this that may enable us to deal with perpetual novelty (see, for example, Holland, 1986). A successful approach combining generating functions, automatic groups, and a revised use of Markov processes should characterize some of the persistent features of the far-from-equilibrium, evolutionary trajectories generated by recombination.

Whatever our mathematical approach to *cas*, the objective remains to determine common causes of common characteristics. When we embarked, I listed three mechanisms—tags, internal models, and building blocks—and four properties—aggregation, nonlinearity, flows, and diversity—that have become the prime candidates for causes and characters in my own search. Other researchers will have other candidates. Nevertheless, at the Santa Fe Institute I think we would all agree on the following broad requirements for a successful approach to theory:

1. *Interdisciplinarity*. Different *cas* show different characteristics of the class to advantage, so that clues come from different *cas* in

different disciplines. In this exposition we've seen many comparisons and the uses to which they can be put.

2. *Computer-based thought experiments.* Computer-based models allow complex explorations not possible with the real system. I have pointed out that it is no more feasible to isolate and repeatedly restart parts of a real *cas* than it is to test flameouts on a real jet airplane carrying passengers. Computer-based models make counterpart experiments possible. Such models can provide existence proofs, which show that given mechanisms are sufficient to generate a given phenomenon. They can also suggest critical patterns and interesting hypotheses to the prepared observer, such as conditions for the existence of lever points.

3. *A correspondence principle.* Bohr's famous principle, translated to *cas*, means that our models should encompass standard models from prior studies in relevant disciplines. Two advantages accrue. Bohr's principle assures relevance of the resulting *cas* theory by requiring it to incorporate hard-won distillations and abstractions from well-established disciplines. It also forestalls what I call "eye of the beholder" errors. Those errors occur when the mapping between a simulation and the phenomena being investigated is insufficiently constrained, allowing the researcher too much freedom in assigning labels to what are, after all, simply number streams in a computer. Standard models from well-established disciplines constrain this freedom because they have been developed with a standard mapping in mind.

4. *A mathematics of competitive processes based on recombination.* Ultimately, we need rigorous generalizations that define the trajectories produced by the interaction of competition and recombination, something computer-based experiments cannot provide on their own. An appropriate mathematics must depart from traditional approaches to emphasize persistent

features of the far-from-equilibrium evolutionary trajectories
generated by recombination.

I believe this amalgam, appropriately compounded, offers hope for a
unified approach to the difficult problems of complex adaptive systems
that stretch our resources and place our world in jeopardy. It is an effort
that can hardly fail. At worst, it will disclose new sights and perspec-
tives. At best, it will reveal the general principles we seek.

Bibliography

(* *Indicates a book or reference accessible to the general reader.*)

* Axelrod, R. 1984. *The Evolution of Cooperation.* New York: Basic Books.
————. 1987. "The Evolution of Strategies in the Iterated Prisoner's Dilemma." In L. D. Davis, ed., *Genetic Algorithms and Simulated Annealing.* Los Altos, Calif.: Morgan Kaufmann.
Baumslag, G. 1994. "Review of *Word Processing in Groups* by D. B. A. Epstein et al." *Bulletin of the American Mathematical Society* 31 (1): 86–91.
Boldrin, M. 1988. "Persistent Oscillations and Chaos in Economic Models: Notes for a Survey." In P. W. Anderson et al., eds., *The Economy as an Evolving Complex System.* Reading, Mass.: Addison-Wesley.
* Bonner, J. T. 1988. *The Evolution of Complexity by Means of Natural Selection.* Princeton: Princeton University Press.
* Brower, L. P., ed. 1988. *Mimicry and the Evolutionary Process.* Chicago: University of Chicago Press.
* Brown, J. H. 1994. "Complex Ecological Systems." in G. A. Cowan et al., eds. *Complexity: Metaphors, Models, and Reality.* Reading, Mass.: Addison-Wesley.
Buss, L. W. 1987. *The Evolution of Individuality.* Princeton: Princeton University Press.
* Dawkins, R. 1976. *The Selfish Gene.* Oxford: Oxford University Press.

Edelman, G. M. 1988. *Topobiology: An Introduction to Molecular Embryology.* New York: Basic Books.

Feller, W. 1950. *An Introduction to Probability Theory and Its Applications.* New York: Wiley.

* Gell-Mann, M. 1994. *The Quark and the Jaguar: Adventures in the Simple and the Complex.* New York: Freeman.

* Gould, S. J. 1994. "The Evolution of Life on Earth." *Scientific American,* October, pp. 84–91.

* Hebb, D. O. 1949. *The Organization of Behavior: A Neuropsychological Theory.* New York: Wiley.

* Hofstadter, D. R. 1979. *Gödel, Escher, Bach: An Eternal Golden Braid.* New York: Basic Books.

Holland, J. H. 1976. "Studies of the Spontaneous Emergence of Self-Replicating Systems Using Cellular Automata and Formal Grammars." In A. Lindenmayer and G. Rozenberg, eds., *Automata, Languages, Development.* Amsterdam: North-Holland.

————. 1986. "A Mathematical Framework for Studying Learning in Classifier Systems." In D. Farmer et al., *Evolution, Games and Learning: Models for Adaptation in Machine and Nature.* Amsterdam: North-Holland.

————. 1992. *Adaptation in Natural and Artificial Systems: An Introductory Analysis with Applications to Biology, Control, and Artificial Intelligence,* 2nd ed. Cambridge, Mass.: MIT Press.

* Hölldobler, B., and E. O. Wilson. 1990. *The Ants.* Cambridge, Mass.: Belknap Press of Harvard University Press.

Kauffman, S. A. 1994. "Whispers from Carnot: The Origins of Order and Principles of Adaptation in Complex Nonequilibrium Systems." In G. A. Cowan et al., eds., *Complexity: Metaphors, Models, and Reality.* Reading, Mass.: Addison-Wesley.

* Lodge, O. 1887 (1950). "Johann Kepler." In J. R. Newman, *The World of Mathematics.* New York: Simon and Schuster.

Lotka, A. J. 1956. *Elements of Mathematical Biology.* New York: Dover.

Marimon, R., E. McGratten, and T. J. Sargent. 1990. "Money as a Medium of Exchange in an Economy with Artificially Intelligent Agents." *Journal of Economic Dynamics and Control* 14: 329–373.

Maynard Smith, J. 1978. *The Evolution of Sex.* Cambridge: Cambridge University Press.

Motowski, A. 1952. *Sentences Undecidable in Formalized Arithmetic: An Exposition of the Theory of Kurt Gödel.* Amsterdam: North-Holland.

* Orel, V. 1984. *Mendel.* Oxford: Oxford University Press.

* Pais, A. 1991. *Niels Bohr's Times: In Physics, Philosophy, and Polity.* Oxford: Oxford University Press.

Perelson, A. S. 1994. "Two Theoretical Problems in Immunology: AIDS and Epitopes." In G. A. Cowan et al., eds., *Complexity: Metaphors, Models, and Reality.* Reading, Mass.: Addison-Wesley.

Perry, Z. A. 1984. "Experimental Study of Speciation in Ecological Niche Theory Using Genetic Algorithms." Doctoral dissertation, University of Michigan.

* Sagan, D., and L. Margulis. 1988. *Garden of Microbial Delights: A Practical Guide to the Subvisible World.* Cambridge, Mass.: Harcourt Brace Jovanovich.

Samuelson, P. A. 1948. *Economics: An Introductory Analysis.* New York: McGraw-Hill.

* Sherrington, C. 1951. *Man on His Nature.* London: Cambridge University Press.

* Smith, A. 1776 (1937). *The Wealth of Nations.* New York: Modern Library.

Srb, A., et al. 1965. *General Genetics.* New York: Freeman.

Turing, A. M. 1937. "On Computable Numbers, with an Application to the Entscheidungsproblem." *Proceedings of the London Mathematical Society,* series 2, no. 4: 230–265.

————. 1952. "The Chemical Basis of Morphogenesis." *Philosophical Transactions of the Royal Society of London,* series B, 237: 37–72.

* Ulam, S. M. 1976. *Adventures of a Mathematician.* New York: Scribners.

von Neumann, J. 1966. *Theory of Self-Reproducing Automata,* ed. A. W. Burks. Urbana: University of Illinois Press.

* Waldrop, M. M. 1992. *Complexity: The Emerging Science at the Edge of Order and Chaos.* New York: Simon and Schuster.

* Weyl, H. 1952. *Symmetry.* Princeton: Princeton University Press.

* Wiener, P. P., ed. 1958. *Values in a Universe of Chance: Selected Writings of Charles S. Peirce.* Garden City, N.Y.: Doubleday.

* Wright, W. 1989. *SimCity* (video game). Orinda, Calif.: Maxis Corporation.

Index